Selected Titles in This Series

718 **Bernhard Lani-Wayda,** Wandering solutions of delay equations with sine-like feedback, 2001
717 **Ron Brown,** Frobenius groups and classical maximal orders, 2001
716 **John H. Palmieri,** Stable homotopy over the Steenrod algebra, 2001
715 **W. N. Everitt and L. Markus,** Multi-interval linear ordinary boundary value problems and complex symplectic algebra, 2001
714 **Earl Berkson, Jean Bourgain, and Aleksander Pełczynski,** Canonical Sobolev projections of weak type $(1,1)$, 2001
713 **Dorina Mitrea, Marius Mitrea, and Michael Taylor,** Layer potentials, the Hodge Laplacian, and global boundary problems in nonsmooth Riemannian manifolds, 2001
712 **Raúl E. Curto and Woo Young Lee,** Joint hyponormality of Toeplitz pairs, 2001
711 **V. G. Kac, C. Martinez, and E. Zelmanov,** Graded simple Jordan superalgebras of growth one, 2001
710 **Brian Marcus and Selim Tuncel,** Resolving Markov chains onto Bernoulli shifts via positive polynomials, 2001
709 **B. V. Rajarama Bhat,** Cocylces of CCR flows, 2001
708 **William M. Kantor and Ákos Seress,** Black box classical groups, 2001
707 **Henning Krause,** The spectrum of a module category, 2001
706 **Jonathan Brundan, Richard Dipper, and Alexander Kleshchev,** Quantum Linear groups and representations of $GL_n(\mathbb{F}_q)$, 2001
705 **I. Moerdijk and J. J. C. Vermeulen,** Proper maps of toposes, 2000
704 **Jeff Hooper, Victor Snaith, and Min van Tran,** The second Chinburg conjecture for quaternion fields, 2000
703 **Erik Guentner, Nigel Higson, and Jody Trout,** Equivariant E-theory for C^*-algebras, 2000
702 **Ilijas Farah,** Analytic guotients: Theory of liftings for quotients over analytic ideals on the integers, 2000
701 **Paul Selick and Jie Wu,** On natural coalgebra decompositions of tensor algebras and loop suspensions, 2000
700 **Vicente Cortés,** A new construction of homogeneous quaternionic manifolds and related geometric structures, 2000
699 **Alexander Fel'shtyn,** Dynamical zeta functions, Nielsen theory and Reidemeister torsion, 2000
698 **Andrew R. Kustin,** Complexes associated to two vectors and a rectangular matrix, 2000
697 **Deguang Han and David R. Larson,** Frames, bases and group representations, 2000
696 **Donald J. Estep, Mats G. Larson, and Roy D. Williams,** Estimating the error of numerical solutions of systems of reaction-diffusion equations, 2000
695 **Vitaly Bergelson and Randall McCutcheon,** An ergodic IP polynomial Szemerédi theorem, 2000
694 **Alberto Bressan, Graziano Crasta, and Benedetto Piccoli,** Well-posedness of the Cauchy problem for $n \times n$ systems of conservation laws, 2000
693 **Doug Pickrell,** Invariant measures for unitary groups associated to Kac-Moody Lie algebras, 2000
692 **Mara D. Neusel,** Inverse invariant theory and Steenrod operations, 2000
691 **Bruce Hughes and Stratos Prassidis,** Control and relaxation over the circle, 2000
690 **Robert Rumely, Chi Fong Lau, and Robert Varley,** Existence of the sectional capacity, 2000
689 **M. A. Dickmann and F. Miraglia,** Special groups: Boolean-theoretic methods in the theory of quadratic forms, 2000

(Continued in the back of this publication)

Wandering Solutions of
Delay Equations with
Sine-Like Feedback

Memoirs
of the
American Mathematical Society

Number 718

Wandering Solutions of
Delay Equations with
Sine-Like Feedback

Bernhard Lani-Wayda

May 2001 • Volume 151 • Number 718 (fourth of 5 numbers) • ISSN 0065-9266

American Mathematical Society
Providence, Rhode Island

2000 *Mathematics Subject Classification.*
Primary 34K15, 58F13, 70K50.

Library of Congress Cataloging-in-Publication Data
Lani-Wayda, Bernhard, 1961–
  Wandering solutions of delay equations with sine-like feedback / Bernhard Lani-Wayda.
    p. cm. — (Memoirs of the American Mathematical Society, ISSN 0065-9266 ; no. 718)
  "Volume 151, number 718 (fourth of 5 numbers)."
  Includes bibliographical references.
  ISBN 0-8218-2680-8
  1. Delay differential equations–Numerical solutions. 2. Attractors (Mathematics) I. Title.
II. Series.
QA3.A57 no. 718
[QA372]
510 s—dc21
[515′.35]   2001018229

Memoirs of the American Mathematical Society

  This journal is devoted entirely to research in pure and applied mathematics.

  Subscription information. The 2001 subscription begins with volume 149 and consists of six mailings, each containing one or more numbers. Subscription prices for 2001 are $494 list, $395 institutional member. A late charge of 10% of the subscription price will be imposed on orders received from nonmembers after January 1 of the subscription year. Subscribers outside the United States and India must pay a postage surcharge of $31; subscribers in India must pay a postage surcharge of $43. Expedited delivery to destinations in North America $35; elsewhere $130. Each number may be ordered separately; *please specify number* when ordering an individual number. For prices and titles of recently released numbers, see the New Publications sections of the *Notices of the American Mathematical Society*.

  Back number information. For back issues see the *AMS Catalog of Publications*.

  Subscriptions and orders should be addressed to the American Mathematical Society, P. O. Box 845904, Boston, MA 02284-5904. *All orders must be accompanied by payment.* Other correspondence should be addressed to Box 6248, Providence, RI 02940-6248.

  Copying and reprinting. Individual readers of this publication, and nonprofit libraries acting for them, are permitted to make fair use of the material, such as to copy a chapter for use in teaching or research. Permission is granted to quote brief passages from this publication in reviews, provided the customary acknowledgment of the source is given.

  Republication, systematic copying, or multiple reproduction of any material in this publication is permitted only under license from the American Mathematical Society. Requests for such permission should be addressed to the Assistant to the Publisher, American Mathematical Society, P. O. Box 6248, Providence, Rhode Island 02940-6248. Requests can also be made by e-mail to reprint-permission@ams.org.

Memoirs of the American Mathematical Society is published bimonthly (each volume consisting usually of more than one number) by the American Mathematical Society at 201 Charles Street, Providence, RI 02904-2294. Periodicals postage paid at Providence, RI. Postmaster: Send address changes to Memoirs, American Mathematical Society, P. O. Box 6248, Providence, RI 02940-6248.

  © 2001 by the American Mathematical Society. All rights reserved.
This publication is indexed in *Science Citation Index*®, *SciSearch*®, *Research Alert*®, *CompuMath Citation Index*®, *Current Contents*®*/Physical, Chemical & Earth Sciences*.
Printed in the United States of America.

  ⊚ The paper used in this book is acid-free and falls within the guidelines
established to ensure permanence and durability.
Visit the AMS home page at URL: http://www.ams.org/

10 9 8 7 6 5 4 3 2 1   06 05 04 03 02 01

Contents

Abstract viii
Notation ix
1. Introduction 1
2. Symbolic dynamics for maps 13
3. Composition of 'local' and 'global' maps 22
4. Linking equations and maps 38
5. Explicit examples 70
6. Appendix (Auxiliary results) 84
References 111
Figures 115

ABSTRACT. We develop a framework for the description of erratic solution behavior in delay equations with sine-like, periodic feedback structure. The mechanism described is associated with heteroclinic solutions joining equilibria with locally positive feedback, and is different from the underlying mechanism in previous examples.

We show that, under a transversality condition, the heteroclinic structure implies the existence of solutions wandering between the different periodic copies of such equilibria in a random manner.

The results apply to an explicitly given, piecewise linear delay equation of the form $\dot{x}(t) = f(x(t-1))$, and to smooth perturbations of this example.

Apart from equations with step functions, these are currently the most explicit examples of 'chaotic' dynamics in scalar delay equations.

1991 *Mathematics Subject Classification.* Primary 34K15, 58F13, 70K50.

Partially supported by the Deutsche Forschungsgemeinschaft (DFG) within the program 'Analysis, Ergodentheorie und Effiziente Simulation Dynamischer Systeme'.
Received by the editor October 15, 1998.

Notation

We introduce some symbols which are used throughout the text. Assume that A and B are subsets of normed spaces E and F.

$E(r)$	Open ball of radius r in E
$\text{int}(A)$	Interior of A
$\text{clos}(A)$	Closure of A
$Lip(A,B)$	Lipschitz continuous functions from A to B
$\text{lip}(f)$	Best possible Lipschitz constant for f
$C^0(A,B)$	Continuous functions from A to B
$BC^0(A,B)$	Bounded and continuous functions from A to B
$BLip(A,B)$	Bounded and Lipschitz continuous functions from A to B
$C^1(A,B)$	Continuously differentiable functions from A to B
$BC^1(A,B)$	Bounded and continuously differentiable functions from A to B with bounded derivative
$BC^1 Lip(A,B)$	Lipschitz continuous functions in $BC^1(A,B)$

Note that boundedness and continuity of Df do not imply Lipschitz continuity of f, if the domain of f is not convex. Note also that if A is compact then $C^j(A,B) = BC^j(A,B)$, $j=0,1$.

On the spaces $BC^0(A,B)$ and $BLip(A,B)$ we use the norm

$$\|f\|_{C^0} := \sup_{x \in A} |f(x)|.$$

On $BC^1(A,B)$ and $BC^1 Lip(A,B)$ we use the norm

$$\|f\|_{C^1} := \max\{\|f\|_{C^0}, \|Df\|_{C^0}\}.$$

We also use symbols like C^1 or BC^1 as adjectives, in statements like 'The map f is BC^1 on the set A'.

For a bounded map f from a set D into a normed space, set

$$\|f\|_\infty := \sup_{x \in D} |f(x)|.$$

If $f : A \to B$ is a function, the graph of f is denoted by $\text{graph } f$. We also denote the subset $\{x + f(x) \mid x \in A\}$ of $E \oplus F$ by $\text{graph } f$.

1. INTRODUCTION

1. Delay equations. Retarded functional differential equations, or delay equations, form a class of mathematical models which allow the system's rate of change to depend on its past history, not only on its present state. Ordinary differential equations are included in this class. The dependence on the past appears naturally in numerous applications in biology, electrical engineering or physiology (see, e.g., [Driver], Chapter V). As a very simple, but typical example, consider the equation

$$(1.1) \qquad \dot{N}(t) = -\mu N(t) + rN(t-h)$$

(see [Diekmann et al.], p.1). It is a model for the time evolution of a population $N(t)$ of adult individuals, with per capita mortality rate $\mu > 0$ and per capita reproduction rate $r > 0$. The delay $h > 0$ expresses the fact that newborns take some time to become adults. In models of electrical circuits, delays may for example appear as a consequence of the presence of capacitive elements which take some time to load up. In model equations for neural networks, delays appear very naturally as transmission times along nerve axons.

Even in the case of a scalar equation (as (1.1)), the introduction of a delay has the dramatic consequence that the state space of the system becomes infinite–dimensional: In order to determine a forward evolution of the system, its state must be prescribed not only for $t = 0$, but for all $t \in [-h, 0]$. It is common practice to normalize the delay to 1, and to take the space

$$C := C^0([-1, 0], \mathbb{R})$$

as state space. (For a system of n scalar equations, one would, of course, take $C := C^0([-1, 0], \mathbb{R}^n)$, but let us focus on scalar equations.) For $t \in \mathbb{R}$ and a continuous function x which is defined, at least, on the interval $[t-1, t]$, the symbol x_t is the usual notation for the *segment* of x at t, that is, the element of C defined by $x_t(s) := x(t+s)$ for all $s \in [-1, 0]$. The class of equations of the form

$$(F) \qquad \dot{x}(t) = F(x_t)$$

with $F : C \to \mathbb{R}$ is a fairly general class of models; in particular, it includes equations of the form

$$\dot{x}(t) = g(x(t)) + f(x(t-1)),$$

with $f, g : \mathbb{R} \to \mathbb{R}$. (Set $F(\psi) := g(\psi(0)) + f(\psi(-1))$ to see this.) Under Lipschitz assumptions on F, existence and uniqueness of *forward solutions* can be proved similarly as for ordinary differential equations. However, for

solutions backward in time, neither existence nor uniqueness can be guaranteed in general. Therefore, such equations do not generate a flow, but only a semiflow on the space C. It assigns to $t \geq 0$ and an initial segment $\psi \in C$ the segment y_t^ψ of the corresponding solution y^ψ at time t.

In this work, the subcategory of equations

$$(f) \qquad \dot{x}(t) = f(x(t-1)),$$

with $f : \mathbb{R} \to \mathbb{R}$, is of particular importance. This class is, on the one hand, too restrictive to yield realistic models for many applications. On the other hand, the infinite dimension of the state space still allows a variety of possible dynamics as rich as one could imagine. (Note the contrast to scalar autonomous ordinary differential equations of first order, where all solutions are monotone.) For example, periodic solutions, connecting solutions between periodic ones, bifurcations, and very complicated dynamical structures are known to occur (see, e.g., [Furumochi], [Nussbaum], [Dormayer 1,2], [Walther 1,2,3]). Even if f is only continuous, a forward solution $x : [-1, \infty) \to \mathbb{R}$ of equation (f) with initial segment $\psi \in C$ is readily obtained by setting $x(t) := \psi(0) + \int_0^t f(\psi(s-1))ds$ for $t \in [0,1]$, then setting $x(t) := x(1) + \int_1^t f(x(s-1))ds$ for $t \in [1,2]$, and so on. Proceeding this way, one obviously gets a handle at studying the dynamics by direct forward computation of solutions. This makes the class of equations (f) particularly well-suited for the study of dynamical mechanisms. (Compare to partial differential equations, where even existence of solutions may be a severe problem.)

Although the models of type (f) will often be too rough to give an accurate description of natural phenomena, it seems likely that they exhibit fundamental mechanisms which occur also in more complicated and realistic equations. Thus the investigation of equation (f) is certainly a reasonable step towards an analytical treatment of more complex equations, which may currently be out of reach. For example, results of this work for equations of type (f) can be extended to equations of type (F) by perturbation arguments.

2. Erratic dynamics in delay and other equations. We mentioned already that the simple form of equation (f) does not exclude complicated dynamics. It was noted by Poincaré, although not proved, that deterministic dynamical systems can contain very irregular motion, today frequently called 'chaos' ([Poincaré]).

One way to describe erratic motions within a dynamical system is to relate them to the index shift operator on a space of symbol sequences; for example, on the space $\{0,1\}^{\mathbb{Z}}$. Symbol shifts are a well-known model for irregular motion, with density of periodic and aperiodic trajectories, dense orbits, and sensitive dependence on initial conditions. Their use in connection with dynamical systems goes back to Hadamard, Birkhoff, and Morse

and Hedlund ([Hadamard], [Birkhoff], [Morse], [Morse, Hedlund]). Poincaré discovered that a *transversally homoclinic orbit* of the return map (named Poincaré map after him) creates complicated geometric structures in phase space, and highly irregular solution behavior. Such an orbit of the Poincaré map arises from a solution of the equations of motion, which converges to a periodic motion in forward and backward time (a homoclinic solution). The global stable and unstable manifolds of the periodic orbit intersect transversally along the homoclinic solution.

Smale proved that the presence of a transversally homoclinic orbit for a map f allows to construct an invariant set in a neighborhood of the homoclinic orbit such that the dynamics of an iterate of f, restricted to this set, is equivalent to a symbol shift ([Smale]). The techniques for the description of transversally homoclinic chaos were extended later to yield a complete description of *all* trajectories in a neighborhood of the homoclinic one ([Palmer], [Kirchgraber, Stoffer]). An extension to noninvertible mappings in infinite dimension was given by Steinlein and Walther, and continued by the author ([Steinlein, Walther], [Lani-Wayda 2]).

The 'transversally homoclinic framework' is mentioned, because all examples of chaos in delay equations known so far fit into this scenario. Some of the earliest numerical observations of irregular dynamical behavior in delay equations were given by Lasota, Wazewska–Czyzewska, Mackey and Glass. The equations were nonlinear variants of equation (1.1) and described physiological control processes ([Lasota], [Lasota, W.-Czyzewska], [Mackey, Glass]).

In a numerical study, Hale and Sternberg detected that the equation studied by Mackey and Glass probably creates homoclinic orbits ([Hale, Sternberg]).

The first analytical proofs for existence of erratic motion in delay equations were given in [Walther 1] and [An der Heiden, Walther], using smoothed step functions as nonlinearities. The fact that these functions are constant on several intervals allows explicit calculation of solutions. It also has the effect that the dynamics in the infinite–dimensional space C essentially collapse to one dimension and can be described by an interval map. Existence of chaotic orbits could then be derived using results from the well-known paper by Li and Yorke [Li, Yorke]. It was remarked later by Hale and Lin that these examples are explicable by transverse homoclinic orbits; an elaborate proof for this fact together with a proof for roughness under perturbations of the chaotic trajectories was later given by the author ([Hale, Lin], [Lani-Wayda 3]).

In the later examples given in [Walther 3], [Lani-Wayda, Walther 2] and [Lani-Wayda 4], solutions transversally homoclinic to unstable periodic solutions were constructed under the premise of certain restrictions on the nonlinearity f in equation (f). The example in [Walther 3] was a sine-like nonlinearity (periodic, with one zero in the interior of the period interval).

The example from [Lani-Wayda, Walther 2], together with the transversality criterion from [Lani-Wayda, Walther 1], showed that transversally homoclinic solutions occur in smooth equations of type (f) with *negative feedback*, i.e., $f(x) \cdot x < 0$ for all $x \neq 0$, and within the class of slowly oscillating solutions. (A related result for step function nonlinearities was given in [Peters].) The well-known Morse decomposition result from [Mallet-Paret] shows that slowly oscillating solutions (which have not more than one zero per time unit) constitute, roughly speaking, the most important part of the dynamics. In [Lani-Wayda 4] it was shown that, under negative feedback, the transversally homoclinic phenomenon can be encountered even if the nonlinearity has only *one* extremum. (For monotone nonlinearities, the results of Mallet-Paret and Sell would exclude erratic solutions [Mallet-Paret, Sell].)

We briefly mention results on irregular dynamics from another branch of infinite–dimensional dynamical systems; namely, parabolic partial differential equations. Poláčik showed that the dynamics of *any* finite-dimensional vector field can be realized on an invariant submanifold of a scalar semilinear parabolic equation (in higher space dimension). In view of known results on ordinary differential equations (e.g., [Kirchgraber, Stoffer]) this means, in particular, that erratic dynamics occur in such parabolic equations ([Poláčik]). Poláčik's result is of fundamental importance because it shows that autonomous parabolic equations in higher space dimensions can exhibit all kinds of behavior, which is not the case in one space dimension. Poláčik's construction does not yield simple analytical formulas for the equations. Here we should also mention the results from [Rybakowski] and from [Faria, Magalhães] on embedding of vector fields into the dynamics of delay equations with several delays, which are in a similar spirit.

Sandstede and Fiedler showed that arbitrary time-periodic vector fields on $\mathbb{R}^2$ can be embedded in the dynamics of nonautonomous parabolic equations with one space dimension (on the circle) [Sandstede, Fiedler]. This approach yields examples in terms of simple formulas.

To the author's knowledge, the above list of known examples for analytically proven existence of 'chaos' in delay equations is complete. We turn to the description of the present paper now.

3. Motivation for this work. Furumochi studied the equation $\dot{x}(t) = \delta + \sin(x(t-L))$ $(L, \delta > 0)$ as a model for control of high frequency oscillators ([Furumochi]). The equation with $\delta = 0$ was used as a model for a similar physical situation in [Wischert et al.].

Using Schauder's fixed point theorem, Furumochi proved existence of periodic solutions of the second kind. (That is, solutions y with $y(t+T) = y(t) + k\omega$ $(t \in \mathbb{R})$ for some $T > 0$, where $k \in \mathbb{Z}$ and $\omega > 0$ is the period of the nonlinearity.)

In a joint paper, Dormayer and the author studied the bifurcation of periodic solutions of the equation

$$(-\alpha \sin) \qquad \dot{x}(t) = -\alpha \sin(x(t-1)),$$

as $\alpha > 0$ is varied ([Dormayer, Lani-Wayda]). It is well-known that equation $(-\alpha \sin)$ has a branch of so-called special symmetric periodic solutions y^α ($\alpha > \pi/2$). These solutions oscillate about zero, have period 4 and amplitude less than π, and the symmetry property $y^\alpha(t) = -y^\alpha(t-2)$ ($t \in \mathbb{R}$) (see, e.g., [Kaplan, Yorke], [Dormayer 2]). Using topological degree theory, it was shown by Walther that secondary bifurcations from this branch occur. Bifurcation of a *smooth* secondary branch of solutions without the above symmetry was proved later by Dormayer ([Walther 2], [Dormayer 1]). In our numerical study, we found successive period doubling bifurcations on the secondary branch. Further, we made the following observation, which is the original motivation for this work:

Solutions which start close to the periodic (nonsymmetric) solutions on the secondary branch oscillate in the interval $(-\pi, \pi)$ for $\alpha < 4.99$. For larger α−values, however, it is typical that numerical solutions start to leave this interval, and to 'wander up and down' between different intervals $((2k-1)\pi, (2k+1)\pi)$ in an unpredictable manner. Similar phenomena were reported in [Wischert et al, p. 211ff].

The purpose of this work is to at least partially explain the experimental observation, by description of a mechanism that creates such behavior for sine-like delay equations. Further, we look for examples given by analytical expressions which are as simple as possible.

As main result, we obtain explicit equations which have solutions wandering between different levels that may be prescribed arbitrarily.

(Note that the solutions of sine-like equations constructed by Walther in the transversally homoclinic setting [Walther 3] do not have this property. Those solutions always go 'up' one level.)

In all of the previous papers on chaos in delay equations listed above, the nonlinearities are either 'artificially' constructed, i.e., defined by properties which are necessary for the proof, or they are smoothed step functions. Most equations appearing in applications are given by smooth (usually analytic) functions. Hence it is certainly desirable to have methods which allow to describe complicated motion in such examples, not only for special constructions or (smoothed) step functions.

Our work is a step in this direction – the primary example that we give is a piecewise linear function, something between step functions and analytical functions, so to speak. By perturbation from this example, we obtain C^1 examples which are piecewise given by explicit analytic expressions. From numerical observations it is very likely that the nonlinearity $f(x) = (1/\pi)[\sin(\pi x) + (\pi/2 - 1)\sin^3(\pi x)]$ exhibits the phenomena described

in this thesis, but we have no proof for this conjecture. Let us now sketch essential ideas of the paper.

4. Basic ideas and methods. It is intuitively clear that strong feedback (large α) is likely to cause solutions to leave the region $(-\pi, \pi)$, where the locally negative feedback of equation $(-\alpha \sin)$ forces solutions to oscillate about zero. However, the following fact should be kept in mind: The result from [Dormayer 2] on the asymptotic behavior of Floquet multipliers as $\alpha \to \infty$ suggests that there exist arbitrarily large α-values where the solutions on the primary branch become stable again. (This conjecture is not proved presently.) Obviously the numerics indicate that, if this conjecture is true, the domains of attraction for these solutions must be 'small'.

We are looking for a structure that makes solutions go up or down between different levels. The numerical simulation for equation $(-\alpha \sin)$ with $\alpha > 4.99$ showed solutions oscillating about $2k\pi$, then wandering to $(2k+2)\pi$, or to $(2k-2)\pi$, oscillating there, and so on ($k \in \mathbb{Z}$). In between the oscillation periods about the 'even' levels $2k\pi$, there sometimes occur short oscillations about the 'odd' levels $(2k \pm 1)\pi$. This behavior suggests to look for heteroclinic solutions that join a periodic solution oscillating about 0 (with values in $(-\pi, \pi)$) to its 2π-translate, which oscillates about 2π (with values in $(\pi, 3\pi)$). Our approach is different, however: We look for heteroclinic solutions that join the *equilibria* at $-\pi$ and π. The reason is simply that such solutions are analytically more accessible, while it seems relatively hopeless to obtain solutions heteroclinic between periodic ones, for explicit nonlinearities. After a normalization which replaces π by 1, we thus obtain a heteroclinic solution x which joins -1 to $+1$. Oddness and 2-periodicity of the nonlinearity imply that all the functions $2k \pm x$ ($k \in \mathbb{Z}$) are also solutions, heteroclinic between $2k \mp 1$ and $2k \pm 1$. We shall see that these solutions provide a transport mechanism that causes the existence of nearby solutions wandering between different integer levels, according to any prescribed level sequence (l_k) with $|l_{k+1} - l_k| = 2$. This property persists under small perturbations of the equation, although such perturbations – in the typical case – destroy the heteroclinic structure.

The dynamical situation is related to the famous Šil'nikov homoclinic phenomenon, which was described for ordinary differential equations in dimension three and higher in [Šil'nikov 1], [Šil'nikov 2]. The classical Šil'nikov situation involves a vector field in $\mathbb{R}^3$, with a spiral saddle at 0, and a homoclinic orbit (compare the upper half of Figure 2). The eigenvalues of the linearization at 0 are of the form $\lambda > 0$ and $\mu = \rho + i\omega$, $\bar{\mu} = \rho - i\omega$, with $\rho < 0 < \omega$. Today, it is well-known that the inequality $\lambda > |\rho|$ implies the presence of erratic motions in the neighborhood of the homoclinic one, which can be described by symbolic dynamics in several ways. For example, it is possible to show that a 2-dimensional return map constructed close to the homoclinic solution contains Smale horseshoes. (These are rectangles which

intersect their images in two components – a structure which is known to contain symbolic dynamics; see also [Guckenheimer, Holmes], [Lani-Wayda 1].) Clearly, the Šil'nikov phenomenon is different from the 'transversely homoclinic solution to a periodic solution'–framework described above, which was appropriate for all examples in delay equations so far.

The situation that we will encounter in delay equations can be visualized in simplified form by the following variant of the Šil'nikov scenario: Consider a family of spiral saddles in $\mathbb{R}^3$ as above, lined up along the z-axis at the values $2k - 1$ ($k \in \mathbb{Z}$), and such that each saddle is connected to the one below *and* to the one above by the lower and upper branch of its unstable manifold. (See Figure 1; the values $2k - 1$ are chosen only for reasons of analogy to the delay equation that we consider later.)

The vector field $v : \mathbb{R}^3 \to \mathbb{R}^3$ should have the symmetries $v(x, y, z + 2) = v(x, y, z) = -v(-x, -y, -z)$. We heuristically explain basic ideas of the construction of solutions that wander between different saddles in an arbitrary manner:

a) Identifying all saddles (which are copies of each other), and shifting them to the origin, one may think of the situation in Figure 1 as only one saddle, with a symmetric pair of homoclinic orbits (see Figure 2; compare, e.g., [Wiggins], p. 253-257). Now we look for solutions that follow the upper or the lower homoclinic orbit in a prescribed sequence.

b) Assume now, for simplification, that the vector field is linear in a neighborhood of 0. (The local stable and unstable manifolds are then given by the xy-plane and the z-axis, respectively.) We use one complex coordinate for the xy-plane. Consider points $p = (w, z)$ ($w \in \mathbb{C}, z \in \mathbb{R}$) with z small. Under the flow $\Phi : \mathbb{R} \times \mathbb{R}^3 \to \mathbb{R}^3$, such points will move close to the origin first, then make a turn along one of the homoclinic orbits, and eventually reach the linear region again. Obviously, it depends only on the sign of z whether the solution starting at p will make a turn along the upper or along the lower branch of the unstable manifold. Fix a number $z_0 > 0$ such that the point $(0, z_0)$ is on the unstable manifold of 0 in the linear region. We assume $z > 0$, and we try to determine the third component z' of the solution through p, when it reaches the linear region again at a large time $T > 0$.

Following the linear vector field, the solution $t \mapsto [\exp(\mu t)w, \exp(\lambda t)z]$ through (w, z) reaches the plane $\mathbb{C} \times \{z_0\}$ at the time $\tau(z) := (1/\lambda) \log(z_0/z)$. It hits that plane at the point $\tilde{p} := (\tilde{w}, z_0)$, where $\tilde{w} = \exp(\mu\tau(z))w = \exp(\rho\tau(z))\exp(i\omega\tau(z))w$.

c) If z is close to 0 then $\tau(z)$ is large and $\tilde{w}$ is close to 0. In this case, the further fate of the point $\tilde{p}$ under the flow is determined by the linearization along the upper homoclinic orbit. More explicitly, $\Phi(T, p) = \Phi(T - \tau(z), \tilde{p})$ can be approximated by $D_2\Phi(T - \tau(z), (0, z_0))[\tilde{p} - (0, z_0)] = D_2\Phi(T - \tau(z), (0, z_0))(\tilde{w}, 0)$.

The linear operator $D := D_2\Phi(T - \tau(z), (0, z_0))$ sends the plane $\mathbb{C} \times \{0\}$

to a space H, and we require

(*) $$H \not\subset \mathbb{C} \times \{0\}.$$

(See the remarks below.)

d) If we denote the projection to the third coordinate by pr_3, it follows that the $\mathbb{R}$–linear functional defined by $\xi(u) := \mathrm{pr}_3 \circ D(u,0)$ ($u \in \mathbb{C}$) is nonzero. In first order approximation, we have

$$z' = \xi(\tilde{w}) = \exp(\rho\tau(z))\xi[\exp(i\omega\tau(z))w].$$

Assume now for simplicity that $\xi(u) = \mathrm{Re}(u)$ ($u \in \mathbb{C}$), and that $w \in \mathbb{R}$. Then

$$\begin{aligned} z' &= \exp(\rho\tau(z)) \cdot \cos(\omega\tau(z)) \cdot w \\ &= w \exp((\rho/\lambda)\log(z_0/z)) \cos[(\omega/\lambda)\log(z_0/z)] \\ &= c_1 z^{-\rho/\lambda} \cos(c_2 \log(z) - c_3), \end{aligned}$$

with appropriate constants c_1, c_2, c_3. Ignoring these constants, we obtain the map defined by

$$z' = f(z) := z^{-\rho/\lambda} \cos(\log z) \qquad (z > 0)$$

as a model for the dependence of z' on the initial value (w, z). In view of the symmetry of the vector field, the odd continuation of this map (for $z < 0$) is then a model for all $z \neq 0$. Observe now that the conditions on the eigenvalues imply that $-\rho/\lambda \in (0, 1)$. It is not difficult to show that the (extended) map f has forward orbits $(z_j)_{j \in \mathbb{N}_0}$ with $\mathrm{sign}(z_j) = s_j$ ($j \in \mathbb{N}_0$), for arbitrary sequences $(s_j) \in \{-1, 1\}^{\mathbb{N}_0}$: One can first construct finite orbit segments following the first k prescribed signs, and then obtain an infinite orbit by a compactness argument. In fact, appropriate initial values are found in every z–interval where $f(z)$ has the sign s_0.

It turns out that the idea sketched very roughly here (with the model map f) can be carried over to delay equations. The infinite-dimensional situation can be thought of as the cross product of the situation from Figure 1 with an infinite–dimensional space. In the linear region, the part of solutions in this space decays at a faster rate than $\exp(\rho t)$ and therefore has no essential influence on the dynamics.

Our construction is geometric and yields geometric information about the location of the initial values for solutions with prescribed behavior. (Geometric arguments in the area of dynamical systems have a tendency to *replace* proofs by figures instead of illustrating them; we hope that this is not the case in the present work.)

Remarks on condition (*): In dimension three, this property is automatically satisfied, because of the following facts (see Figure 2):

(i) D is an invertible linear operator; hence $\dim H = 2$.

(ii) The plane $\mathbb{C} \times \{0\}$ does not contain the tangent direction to the homoclinic orbit at $(0, z_0)$, which is $\{0\} \times \mathbb{R}$. This direction is mapped by D to the direction tangent to the homoclinic orbit at the point $\Phi(T - \tau(z), (0, z_0))$, which is contained in the stable space $\mathbb{C} \times \{0\}$. Hence H does not contain one direction in the stable space $\mathbb{C} \times \{0\}$, and therefore $\dim[H \cap (\mathbb{C} \times \{0\})] = 1$.

(iii) $\dim(H + \mathbb{C} \times \{0\}) = \dim H + 2 - \dim[H \cap (\mathbb{C} \times \{0\})] = 2 + 2 - 1 = 3$.

Now consider an analogous situation in dimension four or higher, still with one unstable eigenvalue, and with a leading pair of complex conjugate eigenvalues $\rho \pm i\omega$ as above (i.e., all other eigenvalues have real part less than ρ). To this pair corresponds a two-dimensional subspace S_2 of the stable space. Then the property analogous to (*) is that D maps S_2 to a space not contained in the stable space. This is a nontrivial additional condition. (We will have to prove a corresponding transversality condition in the infinite–dimensional situation of delay equations.)

So far the description of the fundamental ideas which allow to conclude the existence of erratic solutions, *assuming* the appropriate heteroclinic structure. The actual realization of this structure in concrete examples requires additional techniques and ideas. The piecewise linear ansatz allows explicit calculations, and the use of series expansions for solutions. In addition, we need some estimates for values of explicitly given functions, which can all be verified independently of computers.

5. Comment on the results and related work. Papers of Walther and Lin also treat solutions of delay equations homoclinic or heteroclinic to equilibria, and the possibly bifurcating solutions, as a parameter is varied. Both references ([Walther 4], [Lin]) use the assumption $|\rho| > \lambda$, under which the homoclinic loop is attractive. (This is reflected in the fact that iteration of the above map f converges to zero if $|\rho| > \lambda$). These papers do not describe erratic motion.

In the doubly homoclinic situation of Figure 2 (in $\mathbb{R}^3$), and under the assumption $|\rho| > \lambda$, Holmes showed the following fact: Solutions approaching the two homoclinic orbits can follow any prescribed sequence of upper and lower turns along these orbits [Holmes].

Let us now quote the following two passages from a paper on partial differential equations ([McLaughlin, Shatah, p.94]), and then discuss our results, with reference to the statements from these quotations.

"Symbol dynamics is very appealing because it demonstrates the existence of chaotic motions which last for all time. However, it has some drawbacks. First, it occurs on a very small set in phase space, which is not shown to be (and is likely not) a stable set. As such, this type of chaos may not be observable."

> *"Finally, the construction of the horseshoe is almost always done for generic abstract models, rather than for a fixed specific dynamical system. To us, this generic situation seems to be a severe limitation of the practice of the method. (Sometimes numerical constructions are used to verify the generic assumptions for specific examples, and thus to overcome this limitation. However, in high dimensional singular situations, such numerical verifications are extremely difficult – if not impossible.)"*

Concerning the first 'drawback', it must be admitted that the solutions that we describe do certainly not fill up open sets of phase space. In fact, they do not even exhibit the type of behavior that seems to be dominant in the numerics. As already remarked, in numerical experiments one sees oscillation about even levels, i.e., levels with locally negative feedback, with short possible periods of oscillation about odd levels (with locally positive feedback). Our solutions, on the contrary, oscillate only about odd levels, with monotone behavior in between two such levels.

Complicated behavior on 'large' subsets of phase space, for dynamical systems given by simple analytical expressions, is difficult to prove even in low finite dimension. (Compare, e.g., [Bennedicks, Carleson 1], [Bennedicks, Carleson 2], [Lazutkin].) The computer experiments from [Dormayer, Lani-Wayda] certainly indicate that erratic motion can take place for 'thick' sets of initial values. To prove this remains as an open problem; we can not give a corresponding analytical result for delay equations.

Consider now the second passage quoted above. It is true that there are many publications which assume global conditions like heteroclinic solutions, without establishing them in specific cases. The present work, initiated by experimental observations, makes a contribution to the small collection of explicit examples.

Although the state space is infinite–dimensional, we are in the fortunate situation that all necessary conditions can be verified analytically, independent of machine calculations. This fact may be of interest, in view of the increasing importance of computer–assisted proofs.

6. Guideline for the reader. We give a short outline of how the necessary preparations are carried out and then combined in the following sections.

In Section 2, we give general conditions for maps to have orbits with prescribed signs of one component. The method used here is a combination of classical horseshoe techniques using contraction and expansion properties and of compactness arguments. The main result, Theorem 2.4, gives sufficient analytical conditions for a map to contain symbolic dynamics. The setup was chosen such that the conditions are relatively easy to verify, and that using the inverse mapping (which may not exist) is avoided.

In Section 3 we study maps of the form $\mathcal{T}_1 \circ \mathcal{T}_0$, where $\mathcal{T}_0$ is given by a linear semiflow and describes the local behavior near the equilibrium, and the 'global' map $\mathcal{T}_1$ should be thought of as a map defined by following solutions close to a homoclinic solution. The most essential ingredients here are the explicit expression for the three 'leading' components of $\mathcal{T}_0$, the faster exponential decay of the remaining modes, and the approximation of $\mathcal{T}_1$ by its linearization.

Maps like $\mathcal{T}_1$ and $\mathcal{T}_0$ appear later in the study of delay equations, and it is shown that the composition $\mathcal{T}_1 \circ \mathcal{T}_0$ fits into the framework of Section 2.

Section 4 establishes the connection between maps and delay equations, assuming the existence of heteroclinic solutions, and a transversality type condition, which is an analogue of condition (*) above. Starting from these assumptions, we construct maps $\mathcal{T}_1$ and $\mathcal{T}_0$ with the properties required in Section 3. The map $\mathcal{T}_1 \circ \mathcal{T}_0$ is then, modulo the period of the nonlinearity, a return map defined on a surface of section near the equilibrium -1. In addition, we give the perturbation statements which are necessary to obtain additional examples from the primary, piecewise linear equation. At the end of Section 4, we briefly comment on the local linearity conditions which are assumed in this section.

In Section 5, we construct an explicit example (and perturbations of it) satisfying the conditions of Section 4. The method relies on the piecewise linear structure which allows explicit calculations. In particular, it is important that the local one–dimensional unstable manifold of the equilibrium -1 is explicitly known for all parameters in a certain range, so that forward calculation of the corresponding solution is possible. The parameter can be adjusted such that this solution hits the stable manifold of the equilibrium $+1$. The transversality condition from Section 4 can then be verified by calculation for this particular parameter. The computations use several estimates on the values of analytic functions, the proof of which is deferred to Section 6.

The appendix, Section 6, gives auxiliary technical results which are used in the proofs. Such are perturbation statements for delay equations, results on series expansions for solutions of linear delay equations, and numerical estimates (based on Taylor expansion) on the values of specific functions.

Altogether, the following sections may be divided in the 'more technical part' (Sections 3, 4, and 6) and the 'more original part' (the relatively abstract Section 2 and, in particular, Section 5).

It is possible to read the main results (Theorem 5.4 and Corollary 5.7) after having read only Definitions 4.1 and 4.2. A reader going deeply into detail might find that several of the technical results are of some interest by themselves, as dynamical systems tools. For example, Theorem 6.7 on Poincaré type maps defined in a neighborhood of a set with more than one

point, or the smoothness statement for piecewise C^1 equations of Lemma 6.5.

A reader familiar with Šil'nikov type results and horseshoe constructions could take the main result of Section 4 (Theorem 4.14) on trust, at first reading. One may then focus on Section 5, where the desired global dynamical properties are shown to be present in a simple example.

Acknowledgements

My personal thanks go to Dr. Hans-Otto Walther for continuous encouragement and support over many years.
Thanks to Peter Dormayer for suggesting the numerical studies which later gave the motivation for this work.
I am indebted to the Deutsche Forschungsgemeinschaft (DFG) and the Alexander-von-Humboldt-Stiftung for support in my postdoctoral period.
Thanks to the referees for suggestion of some final improvements.
Special thanks to my wife Brigitte Brink for her understanding.

2. Symbolic dynamics for maps

In this section we consider maps defined on subsets of $M \times \mathbb{R}$, where M is some set. We describe orbits of such maps by encoding the sign of the second component of the points on the orbit. Thus we obtain symbol sequences which are made up of the two symbols $+1$ and -1.

If χ is a map defined on a set D, we denote by $\text{inv}^+(\chi, D)$ the set of all points $x \in D$ with the property that all iterates $\chi^j(x)$ ($j \in \mathbb{N}_0$) are defined and contained in D. Further, we define $\text{traj}(\chi, D)$ as the set of sequences $(x_j)_{j \in \mathbb{Z}}$ of points in D with $x_j = \chi(x_{j-1})$ for all $j \in \mathbb{Z}$.

We use the following two spaces of symbol sequences:

$$\mathcal{S}^+ := \{-1, 1\}^{\mathbb{N}_0} = \{\mathbf{s} = (s_0, s_1, s_2 ...) \mid s_i \in \{-1, 1\} \ (i \in \mathbb{N}_0)\},$$
$$\mathcal{S} := \{-1, 1\}^{\mathbb{Z}} = \{\mathbf{s} = (...s_{-1}, s_0, s_1, ...) \mid s_i \in \{-1, 1\} \ (i \in \mathbb{Z})\}.$$

Assume now that $D \subset M \times (\mathbb{R} \setminus \{0\})$ and $\chi : D \to M \times \mathbb{R}$ is a map. Let $\text{pr}_2 : M \times \mathbb{R} \to \mathbb{R}$ denote the projection. We can define two 'symbolization' maps

$$\sigma^+ : \text{inv}^+(\chi, D) \to \mathcal{S}^+, \quad x \mapsto \mathbf{s} = (s_0, s_1, ...),$$

where $s_j := \text{sign} \, \text{pr}_2 \chi^j(x)$ ($j \in \mathbb{N}_0$), and

$$\sigma : \text{traj}(\chi, D) \to \mathcal{S}, \quad (x_j)_{j \in \mathbb{Z}} \mapsto (\text{sign} \, \text{pr}_2 \, x_j)_{j \in \mathbb{Z}}.$$

(We do not denote the dependence of σ on the domain D.)

2.1. Definition. *We say that χ has symbolic dynamics with respect to $\mathcal{S}^+$ on a subset $\Delta \subset D$ if the map $\sigma^+ : \text{inv}^+(\chi, \Delta) \to \mathcal{S}^+$ is surjective. Similarly, we say that χ has symbolic dynamics with respect to $\mathcal{S}$ on the set Δ if the map $\sigma : \text{traj}(\chi, \Delta) \to \mathcal{S}$ is surjective.*

The above definition expresses existence of many qualitatively different orbits of χ. It is trivial that symbolic dynamics with respect to $\mathcal{S}$ implies the same with respect to $\mathcal{S}^+$. Under compactness assumptions, the reverse implication holds:

2.2. Proposition. *Assume that M is a metric space, $D \subset M \times (\mathbb{R} \setminus \{0\})$, and that $\chi : D \to M \times \mathbb{R}$ is continuous. Assume that $\Delta \subset D$ is closed in $M \times \mathbb{R}$, and that $\text{clos}(\chi(\Delta))$ is compact. Then, if χ has symbolic dynamics with respect to $\mathcal{S}^+$ on Δ, the same also holds with respect to $\mathcal{S}$ on Δ.*

Proof. Let $\mathbf{s} = (...s_{-1} s_0 s_1...) \in \mathcal{S}$. For $j \in \mathbb{N}_0$, define $\mathbf{s}^{(j)} = (s_0^{(j)}, s_1^{(j)}, ...) \in \mathcal{S}^+$ by $s_k^{(j)} = s_{-j+k}$ ($k \in \mathbb{N}_0$). Since χ has symbolic dynamics with respect to $\mathcal{S}^+$ on Δ, for every $j \in \mathbb{N}_0$ there exists $x^j \in \Delta$ with $\sigma^+(x^j) = \mathbf{s}^{(j)}$. In particular,

$$\chi^j(x^j) \in \text{inv}^+(\chi, \Delta), \text{ and } \sigma^+(\chi^j(x^j)) = \mathbf{s}^{(0)} \ (j \in \mathbb{N}_0).$$

We have $\chi^j(x^j) \in \chi(\Delta) \cap \mathrm{inv}^+(\chi, \Delta) \subset \mathrm{clos}(\chi(\Delta)) \cap \mathrm{inv}^+(\chi, \Delta)$, and the latter set is compact, since $\mathrm{clos}(\chi(\Delta))$ is compact and since $\mathrm{inv}^+(\chi, \Delta) = \bigcap_{k \geq 0} \chi^{-k}(\Delta)$ is closed, because Δ is closed. Thus there exist a subsequence $(x^{\varphi^0(j)})_{j \in \mathbb{N}_0}$ of $(x^j)_{j \in \mathbb{N}_0}$ and $z_0 \in \mathrm{clos}(\chi(\Delta)) \cap \mathrm{inv}^+(\chi, \Delta)$ with

$$\chi^{\varphi^0(j)}(x^{\varphi^0(j)}) \to z_0 \quad (j \to \infty).$$

Recursively, we can choose subsequences $(x^{\varphi^J(j)})_{j \in \mathbb{N}_0}$ of $(x^{\varphi^0(j)})_{j \in \mathbb{N}_0}$, for $J \in \mathbb{N}$, such that
   a) $(x^{\varphi^{J+1}(j)})_{j \in \mathbb{N}_0}$ is a subsequence of $(x^{\varphi^J(j)})_{j \in \mathbb{N}_0}$ for $J \in \mathbb{N}_0$, and
   b) For all $J \in \mathbb{N}_0$, there exists $z_{-J} \in \mathrm{clos}(\chi(\Delta)) \cap \mathrm{inv}^+(\chi, \Delta)$ with

$$\chi^{\varphi^J(j) - J}(x^{\varphi^J(j)}) \to z_{-J} \quad (j \to \infty).$$

Define $z_J := \chi^J(z_0)$ for $J \in \mathbb{N}$. It suffices to prove the following statement:
Claim: $(z_J)_{J \in \mathbb{Z}} \in \mathrm{traj}(\chi, \Delta)$, and $\sigma((z_J)_{J \in \mathbb{Z}}) = \mathbf{s}$.
Proof. Let $J \in \mathbb{N}_0$. We know that

$$\chi^{\varphi^J(j) - J}(x^{\varphi^J(j)}) \to z_{-J} \quad (j \to \infty), \text{ and}$$
$$\chi^{\varphi^{J+1}(j) - (J+1)}(x^{\varphi^{J+1}(j)}) \to z_{-(J+1)} \quad (j \to \infty).$$

Since χ is continuous, we have

$$\chi(z_{-J-1}) = \lim_{j \to \infty} \chi[\chi^{\varphi^{J+1}(j) - (J+1)}(x^{\varphi^{J+1}(j)})]$$
$$= \lim_{j \to \infty} \chi^{\varphi^{J+1}(j) - J}(x^{\varphi^{J+1}(j)})$$

(i.e., the last limit exists and equals $\chi(z_{-J-1})$). Now, since $(x^{\varphi^{J+1}(j)})_{j \in \mathbb{N}_0}$ is a subsequence of $x^{\varphi^J(j)}$, the sequence $[\chi^{\varphi^{J+1}(j) - J}(x^{\varphi^{J+1}(j)})]_{j \in \mathbb{N}_0}$ is a subsequence of the sequence $[\chi^{\varphi^J(j) - J}(x^{\varphi^J(j)})]_{j \in \mathbb{N}_0}$, which converges to z_{-J}. Thus the last limit equals z_{-J}, and we obtain

$$z_{-J} = \chi(z_{-J-1}) \text{ for } J \in \mathbb{N}_0.$$

It is clear that $\chi(z_{J-1}) = z_J$ for $J \in \mathbb{N}$, and so

$$(z_J)_{J \in \mathbb{Z}} \in \mathrm{traj}(f, \Delta).$$

Note that $\Delta \subset D$ implies $\Delta \cap (M \times \{0\}) = \emptyset$, so the numbers $\mathrm{sign}\,\mathrm{pr}_2(z_J)$ are defined for $J \in \mathbb{Z}$, and it remains to show the following statement:

(2.2.1) $$\mathrm{sign}\,\mathrm{pr}_2(z_J) = s_J \text{ for } J \in \mathbb{Z}.$$

Since $\sigma^+(x^j) = \mathbf{s}^{(j)}$ for all $j \in \mathbb{N}_0$, we have

$$\sigma^+(x^{\varphi^J(j)}) = \mathbf{s}^{(\varphi^J(j))} \text{ for } J, j \in \mathbb{N}_0.$$

Therefore,

$$\operatorname{sign} \operatorname{pr}_2 \chi^{\varphi^J(j)-J}(x^{\varphi^J(j)}) = s^{(\varphi^J(j))}_{\varphi^J(j)-J} = s_{-J} \text{ for } J, j \in \mathbb{N}_0.$$

Letting $j \to \infty$, it follows that

$$\operatorname{sign} \operatorname{pr}_2 z_{-J} = s_{-J} \text{ for } J \in \mathbb{N}_0.$$

Since $\sigma^+(\chi^j(x^j)) = \mathbf{s}^{(0)}$ for all $j \in \mathbb{N}_0$, we also have

$$\sigma^+[\chi^{\varphi^0(j)}(x^{\varphi^0(j)})] = \mathbf{s}^{(0)} \text{ for all } j \in \mathbb{N}_0,$$

and thus

$$\operatorname{sign} \operatorname{pr}_2 \chi^J[\chi^{\varphi^0(j)}(x^{\varphi^0(j)})] = \mathbf{s}^{(0)}_J = s_J \text{ for } J \in \mathbb{N}_0.$$

Recalling the definition of z_0, and of z_J for $J \in \mathbb{N}_0$, and letting $j \to \infty$, we see that $\operatorname{sign} \operatorname{pr}_2(z_J) = s_J$ for $J \in \mathbb{N}_0$. Property (2.2.1) is proved. $\square$

The following statement is trivial but convenient for later application.

2.3. Remark. *Let M and $\tilde{M}$ be sets, $D \subset M \times (\mathbb{R} \setminus \{0\})$ and $\tilde{D} \subset \tilde{M} \times (\mathbb{R} \setminus \{0\})$. Let a map $\kappa : D \to \tilde{M} \times \mathbb{R}$ with $\kappa(D) \subset \tilde{D}$ and maps $\chi : D \to M \times \mathbb{R}$, $\tilde{\chi} : \tilde{D} \to \tilde{M} \times \mathbb{R}$ be given such that*

$$\kappa \circ \chi = \tilde{\chi} \circ \kappa \text{ on } \chi^{-1}(D).$$

Assume that κ has the property

(2.3.1) $\qquad\qquad \operatorname{sign} pr_2 x = \operatorname{sign} pr_2 \kappa(x) \quad (x \in D).$

Then the following implications hold for any subset $\Delta \subset D$.

(2.3.2) $\quad x \in \operatorname{inv}^+(\chi, \Delta) \Longrightarrow \kappa(x) \in \operatorname{inv}^+(\tilde{\chi}, \kappa(\Delta))$, and $\sigma^+(x) = \sigma^+(\kappa(x))$.

(2.3.3) $\quad\begin{aligned}&(x_j)_{j \in \mathbb{Z}} \in \operatorname{traj}(\chi, \Delta) \Longrightarrow \\ &(\kappa(x_j))_{j \in \mathbb{Z}} \in \operatorname{traj}(\tilde{\chi}, \kappa(\Delta)), \quad \sigma[(x_j)_{j \in \mathbb{Z}}] = \sigma[(\kappa(x_j))_{j \in \mathbb{Z}}].\end{aligned}$

If χ has symbolic dynamics with respect to $\mathcal{S}^+$ or to $\mathcal{S}$ on Δ, then $\tilde{\chi}$ has symbolic dynamics with respect to the same set on $\kappa(\Delta)$.

Proof. Let $x \in \operatorname{inv}^+(\chi, \Delta)$, and set $\mathbf{s} = (s_0, s_1, \ldots) := \sigma^+(x)$. Define $y_j := \kappa(\chi^j(x)) \in \kappa(\Delta)$ $(j \in \mathbb{N}_0)$. Since $\chi^j(x) \in \chi^{-1}(D)$ for all $j \in \mathbb{N}_0$, we have

$$y_{j+1} = (\kappa \circ \chi)(\chi^j(x)) = (\tilde{\chi} \circ \kappa)(\chi^j(x)) = \tilde{\chi}(y_j) \quad (j \in \mathbb{N}_0).$$

Thus $y_0 = \kappa(x) \in \mathrm{inv}^+(\tilde{\chi}, \kappa(\Delta))$, and property (2.3.1) implies that

$$\mathrm{sign}\,\mathrm{pr}_2 \tilde{\chi}^j(y_0) = \mathrm{sign}\,\mathrm{pr}_2 y_j = \mathrm{sign}\,\mathrm{pr}_2 \chi^j(x) = s_j \quad (j \in \mathbb{N}_0),$$

so $\sigma^+(y_0) = \mathbf{s} = \sigma^+(x)$, and (2.3.2) is proved. The proof of (2.3.3) is analogous, and the last assertion follows immediately from (2.3.2) and (2.3.3). □

In the remainder of this section, we give explicit sufficient conditions for the existence of symbolic dynamics. We will see later how these conditions can be verified for concrete examples of maps that describe the dynamics of delay equations. The theorem that we prove here is related to the Smale horseshoe theorem (see, e.g., [Moser]); essentially it gives a weaker conclusion under weaker assumptions. The techniques of the proof are partially similar to those from the author's diploma thesis [Lani-Wayda 1]. Theorem 2.4 is designed to be easily applicable, and stated such that the stability of the conclusions under small perturbations is immediately obvious.

It is possible to prove existence of symbolic dynamics for horseshoe–like maps under more general conditions, using topological methods (the Conley index theory; see e.g. [Mischaikow, Mrozek]). These methods do not provide geometric information about the location of points which follow a symbol sequence under iteration. The analytical conditions of Theorem 2.4 below allow to prove that, for forward symbol sequences, such points lie on Lipschitz graphs of codimension one.

Let S be a normed space and let $I = I_1 \cup I_{-1} \subset S$ be the union of two closed convex sets. We assume that either $I_1 = I_{-1}$ or $\mathrm{dist}(I_1, I_{-1}) > 0$. For $i = -1, 1$ let $J_{i,1} = [a_{i,1}, b_{i,1}]$ be intervals with $0 < a_{i,1} < b_{i,1}$. Similarly, let $J_{i,-1} = [b_{i,-1}, a_{i,-1}]$ be intervals with $b_{i,-1} < a_{i,-1} < 0$. In case $I_1 = I_{-1}$, we assume $J_{1,j} = J_{-1,j}$ ($j = -1, 1$). Let

$$\Delta := \bigcup_{i,j=-1,1} (I_i \times J_{i,j}),$$

and let D be an open neighborhood of Δ in $S \times \mathbb{R}$ (see Figure 3).

Let $\mathrm{pr}_1 : S \times \mathbb{R} \to S$ and $\mathrm{pr}_2 : S \times \mathbb{R} \to \mathbb{R}$ denote the projections. If A is a subset of a metric space and $d > 0$, we denote by $\mathrm{int}_d(A)$ the set of all points $a \in A$ with the property that the open ball of radius d around a is contained in A. (We use the symbol $|\ |$ for the norms on several spaces.)

2.4. Theorem. *Assume that $\chi : D \to S \times \mathbb{R}$ is a C^1 map and that there exist $L > 0$ and $d > 0$ such that the following conditions hold.*

(i) $\forall (x, z) \in \Delta$:

$$z > 0 \implies \mathrm{pr}_1 \chi(x, z) \in \mathrm{int}_d(I_1),$$
$$z < 0 \implies \mathrm{pr}_1 \chi(x, z) \in \mathrm{int}_d(I_{-1}).$$

(ii) *With $B := \max_{i,j=-1,1} |b_{i,j}|$ one has for $i, j = -1, 1$ and $x \in I_i$:*

$$|\mathrm{pr}_2 \chi(x, a_{i,j})| > B, \quad |\mathrm{pr}_2 \chi(x, b_{i,j})| > B, \text{ and}$$
$$\mathrm{sign}\, \mathrm{pr}_2 \chi(x, a_{i,j}) = -\mathrm{sign}\, \mathrm{pr}_2 \chi(x, b_{i,j}).$$

(iii) *One has* $\inf_{(x,z) \in \Delta} |D_2 \mathrm{pr}_2 \chi(x, z)| - L |D_2 \mathrm{pr}_1 \chi(x, z)| > 0$, *and*

$$\sup_{(x,z) \in \Delta} \frac{|D_1 \mathrm{pr}_2 \chi(x, z)| + L |D_1 \mathrm{pr}_1 \chi(x, z)|}{|D_2 \mathrm{pr}_2 \chi(x, z)| - L |D_2 \mathrm{pr}_1 \chi(x, z)|} < L.$$

Then, for every symbol sequence $\mathbf{s} = (s_0, s_1, ...) \in \{-1, 1\}^{\mathbb{N}_0}$, there exists a function $\varphi_{\mathbf{s}} : I_1 \cup I_{-1} \to \mathbb{R}$ with the subsequent properties.

a) *graph $\varphi_{\mathbf{s}} \subset (I_1 \times J_{1,1}) \cup (I_{-1} \times J_{-1,1})$ in case $s_0 = 1$, and graph $\varphi_s \subset (I_1 \times J_{1,-1}) \cup (I_{-1} \times J_{-1,-1})$ if $s_0 = -1$.*
b) *The restrictions $\varphi_{\mathbf{s}}|_{I_1}$ and $\varphi_{\mathbf{s}}|_{I_{-1}}$ have Lipschitz constant L.*
c) *graph $\varphi_{\mathbf{s}} \subset \mathrm{inv}^+(\chi, \Delta)$, and for all $(x, z) \in \mathrm{graph}\, \varphi_{\mathbf{s}}$ one has*

$$\sigma^+(x, z) = \mathbf{s}.$$

In particular, χ has symbolic dynamics with respect to $\mathcal{S}^+$ on the set Δ.

Proof. (See also Figure 4.) Set $R_1 := I_1 \times J_{1,1} \cup I_{-1} \times J_{-1,1}$, and $R_{-1} := I_1 \times J_{1,-1} \cup I_{-1} \times J_{-1,-1}$. For $k \in \mathbb{N}_0$ and $\mathbf{s} = (s_j)_{j \in \mathbb{N}_0} \in \{-1, 1\}^{\mathbb{N}_0}$, define

$$D_{k,\mathbf{s}} := \{(x, z) \in D \mid \chi^j(x, z) \text{ is defined for } j \in \{0, ..., k\},$$
$$\text{and } \chi^j(x, z) \in R_{s_j} (j = 0, ..., k)\}.$$

Since R_1 and R_{-1} are closed in $I \times \mathbb{R}$, the set $D_{k,\mathbf{s}} = \bigcap_{j=0}^{k} \chi^{-j}(R_{s_j})$ is closed in $I \times \mathbb{R}$ for all $k \in \mathbb{N}_0$.

Claim: For all $\mathbf{s} \in \{-1, 1\}^{\mathbb{N}_0}$ and all $k \in \mathbb{N}_0$, there exists an open neighborhood $I_{k,\mathbf{s}}$ of I in S and a C^1-function $\varphi_{k,\mathbf{s}} : I_{k,\mathbf{s}} \to \mathbb{R}$ such that

$$\forall x \in I : |D\varphi_{k,\mathbf{s}}(x)| \leq L, \quad \text{and graph } \varphi_{k,\mathbf{s}}|_I \subset D_{k,\mathbf{s}}.$$

Proof. (By induction on k).

$\underline{k = 0}$: Let $\mathbf{s} \in \{1, -1\}^{\mathbb{N}_0}$. Define $I_{0,\mathbf{s}}$ as an open neighborhood $\tilde{I}_1$ of I, if $I_1 = I_{-1}$, and define $I_{0,\mathbf{s}}$ as the union of two disjoint open neighborhoods $\tilde{I}_1$ and $\tilde{I}_{-1}$ of I_1, respectively of I_{-1}, if $\mathrm{dist}(I_1, I_{-1}) > 0$. Define

$$\varphi_{0,\mathbf{s}}(x) := \begin{cases} \frac{1}{2}(a_{-1,1} + b_{-1,1}) \text{ for } x \in \tilde{I}_{-1}, \\ \frac{1}{2}(a_{1,1} + b_{1,1}) \text{ for } x \in \tilde{I}_1 \end{cases}$$

if $s_0 = 1$, and

$$\varphi_{0,\mathbf{s}}(x) := \begin{cases} \frac{1}{2}(a_{-1,-1} + b_{-1,-1}) \text{ for } x \in \tilde{I}_{-1}, \\ \frac{1}{2}(a_{1,-1} + b_{1,-1}) \text{ for } x \in \tilde{I}_1 \end{cases}$$

if $s_0 = -1$. Then $D\varphi_{0,\mathbf{s}} = 0$ and graph $\varphi_{0,\mathbf{s}} \subset D_{0,\mathbf{s}} = R_{s_0}$.

$\underline{k \to k+1}$: Let $\mathbf{s} = (s_0, s_1, \ldots) \in \{1, -1\}^{\mathbb{N}_0}$. Define $\mathbf{t} = (t_j)_{j \in \mathbb{N}_0} \in \{1, -1\}^{\mathbb{N}_0}$ by $t_j := s_{j+1}$, $j \in \mathbb{N}_0$. Then $(t_0, \ldots, t_k) = (s_1, \ldots, s_{k+1})$. Note that the definitions imply

(2.4.1) $$R_{s_0} \cap \chi^{-1}(D_{k,\mathbf{t}}) = D_{k+1,\mathbf{s}}.$$

From the induction hypotheses, there exists an open neighborhood $N_{k,\mathbf{t}}$ of I in S and a C^1 function $\varphi_{k,\mathbf{t}} : N_{k,\mathbf{t}} \to \mathbb{R}$ with $\|D\varphi_{k,\mathbf{t}}\|_\infty \leq L$ and such that graph $\varphi_{k,\mathbf{t}}\big|_I \subset D_{k,\mathbf{t}}$; in particular, $\|\varphi_{k,\mathbf{t}}\big|_I\|_\infty \leq B$.

Consider the case $s_0 = 1$. Continuity of $\mathrm{pr}_1 \circ \chi$ and condition (i) imply that there exists an open neighborhood W of R_1 in $S \times \mathbb{R}$ such that $W \subset D$ and such that
$$\forall (x, z) \in W: \quad \mathrm{pr}_1 \chi(x, z) \in N_{k,\mathbf{t}}.$$

The function $h : W \to \mathbb{R}$,

$$h(x, z) := \mathrm{pr}_2 \chi(x, z) - \varphi_{k,\mathbf{t}}(\mathrm{pr}_1 \chi(x, z))$$

is well-defined and of class C^1. For $j \in \{-1, 1\}$, $x \in I_j$ and $z \in J_{j,1}$, one obviously has the equivalence

(2.4.2) $$h(x, z) = 0 \iff \chi(x, z) \in \mathrm{graph}\, \varphi_{k,\mathbf{t}}\big|_I.$$

It follows from condition (ii), together with $\|\varphi_{k,\mathbf{t}}\big|_I\|_\infty \leq B$, that

(2.4.3) $$\mathrm{sign}\, h(x, a_{j,1}) = -\mathrm{sign}\, h(x, b_{j,1}) \text{ if } x \in I_j, j \in \{-1, 1\}.$$

For $(x, z) \in R_1 \subset \Delta$ we obtain, using condition (iii) and the estimate $\|D\varphi_{k,\mathbf{t}}\|_\infty \leq L$, that

(2.4.4)
$$\begin{aligned}|D_2 h(x,z)| &= |D_2\mathrm{pr}_2 \mathcal{X}(x,z) - D\varphi_{k,\mathbf{t}}(...)D_2\mathrm{pr}_1\mathcal{X}(x,z)| \\ &\geq |D_2\mathrm{pr}_2\mathcal{X}(x,z)| - \|D\varphi_{k,\mathbf{t}}\|_\infty \cdot |D_2\mathrm{pr}_1\mathcal{X}(x,z)| \\ &\geq |D_2\mathrm{pr}_2\mathcal{X}(x,z)| - L|D_2\mathrm{pr}_1\mathcal{X}(x,z)| \\ &> 0.\end{aligned}$$

Continuity of h, and uniform continuity of $D_2 h$ on the compact sets $\{x\} \times J_{j,1}$ ($x \in I_j$, $j \in \{-1,1\}$) imply the following fact: There exist open neighborhoods I'_j of I_j in S ($j = -1, 1$) such that the set $R'_1 := I'_1 \times J_{1,1} \cup I'_{-1} \times J_{-1,1}$ satisfies $R'_1 \subset W$, and that we have

(2.4.3') $\qquad \mathrm{sign}\, h(x, a_{j,1}) = -\mathrm{sign}\, h(x, b_{j,1})$ if $x \in I'_j$, $j \in \{-1, 1\}$,

(2.4.4') $\qquad \forall\, (x,z) \in R'_1:\quad |D_2 h(x,z)| > 0.$

Properties (2.4.3') and (2.4.4') imply that for $x \in I'_j$ ($j = -1, 1$) there exists a unique $z_x \in \mathrm{int}(J_{j,1})$ such that $h(x, z_x) = 0$. Set $I_{k+1,\mathbf{s}} := I'_1 \cup I'_{-1}$. It follows from (2.4.4'), the Implicit Function Theorem, and the uniqueness of the z_x that the map $\varphi_{k+1,\mathbf{s}} : I_{k+1,\mathbf{s}} \ni x \mapsto z_x$ is C^1. Further, for $x \in I$, one has $(x, z_x) = (x, \varphi_{k+1,\mathbf{s}}(x)) \in \Delta$, and the derivative formula from the Implicit Function Theorem together with (2.4.4), condition (iii), and with the induction hypotheses shows that

$$\begin{aligned}&|D\varphi_{k+1,\mathbf{s}}(x)| \\ &= |-(D_2 h(x, z_x))^{-1}[D_1\mathrm{pr}_2\mathcal{X}(x,z_x) - D\varphi_{k,\mathbf{t}}(\mathrm{pr}_1\mathcal{X}(x,z_x))D_1\mathrm{pr}_1\mathcal{X}(x,z_x)]| \\ &\leq \frac{|D_1\mathrm{pr}_2\mathcal{X}(x,z_x))| + L|D_1\mathrm{pr}_1\mathcal{X}(x,z_x)|}{|D_2\mathrm{pr}_2\mathcal{X}(x,z_x)| - L|D_2\mathrm{pr}_1\mathcal{X}(x,z_x)|} < L.\end{aligned}$$

It is clear that $\mathrm{graph}\, \varphi_{k+1,\mathbf{s}}\big|_I \subset R_1$. We see from (2.4.2) and the induction hypothesis that

$$\mathrm{graph}\, \varphi_{k+1,\mathbf{s}}\big|_I \subset \mathcal{X}^{-1}(\mathrm{graph}\, \varphi_{k,\mathbf{t}}\big|_I) \subset \mathcal{X}^{-1}(D_{k,\mathbf{t}}).$$

In view of (2.4.1) we conclude that

$$\mathrm{graph}\, \varphi_{k+1,\mathbf{s}}\big|_I \subset R_1 \cap \mathcal{X}^{-1}(D_{k,\mathbf{t}}) = D_{k+1,\mathbf{s}}.$$

The claim is proved in case $s_0 = 1$, and the case $s_0 = -1$ is analogous, with $J_{j,1}$ replaced by $J_{j,-1}, j = -1, 1$, and R_1 replaced by R_{-1}.

Let now $\mathbf{s} = (s_0, s_1, ...) \in \{-1, 1\}^{\mathbb{N}_0}$ and choose a sequence $(\varphi_{k,\mathbf{s}})_{k \in \mathbb{N}_0}$ as guaranteed by the above claim. For $j \in \{-1, 1\}$, convexity of I_j and $\|D\varphi_{k,\mathbf{s}}\|_\infty \leq L$ imply that $\varphi_{k,\mathbf{s}}|_{I_j}$ has Lipschitz constant L, for all $k \in \mathbb{N}_0$.

From Remark 2.6 below, applied to $M := I_j$ and $\varphi_{k,\mathbf{s}}|_{I_j}$, we obtain that the function

$$\varphi_\mathbf{s} : I \to \mathbb{R}, \quad \varphi_\mathbf{s}(x) := \lim_{k \to \infty} \sup \varphi_{k,\mathbf{s}}(x)$$

has Lipschitz constant L when restricted to I_j ($j = -1, 1$). Property b) is proved. From Remark 2.7 below, applied to $M := I$, to $(\varphi_{k,\mathbf{s}}|_I)_{k \in \mathbb{N}_0}$, and with $D_k := D_{k,\mathbf{s}}$, we obtain

(2.4.5) $$\operatorname{graph} \varphi_\mathbf{s} \subset \bigcap_{k \in \mathbb{N}_0} D_{k,\mathbf{s}}.$$

In particular, since $D_{0,\mathbf{s}} = R_{s_0} = (I_1 \times J_{1,s_0}) \cup (I_{-1} \times J_{-1,s_0})$, we see that $\varphi_\mathbf{s}$ has property a).

Proof of property c): For $(x, z) \in \operatorname{graph} \varphi_\mathbf{s}$, the definition of the sets $D_{k,\mathbf{s}}$ and (2.4.5) show that $\chi^j(x, z)$ is defined and contained in R_{s_j} for all $j \in \mathbb{N}_0$. Since $R_{s_j} \subset I \times \mathbb{R}^+$ if $s_j = 1$, and $R_{s_j} \subset I \times \mathbb{R}^-$ if $s_j = -1$, we see that $\sigma^+(x, z) = \mathbf{s}$. The statement about symbolic dynamics is clear. □

2.5. Corollary. *In the situation of Theorem 2.4, there exists $\gamma_1 > 0$ with the following property. If $\overline{D} \subset D$ is an open neighborhood of Δ in $S \times \mathbb{R}$, and if $\overline{\chi} \in C^1(\overline{D}, S \times \mathbb{R})$ is such that*

$$\overline{\chi} - \chi_{|\overline{D}} \in BC^1(\overline{D}, S \times \mathbb{R}), \text{ and } \|\overline{\chi} - \chi_{|\overline{D}}\|_{C^1} \leq \gamma_1,$$

then the assertions of Theorem 2.4 also hold for $\overline{\chi}$ instead of χ.

Proof. In view of the strict inequalities in conditions (ii) and (iii) from Theorem 2.4, there exists $\tilde{\gamma}_1 > 0$ such that if $\|\overline{\chi} - \chi_{|\overline{D}}\|_{C^1} \leq \tilde{\gamma}_1$ then $\overline{\chi}$ satisfies these conditions. If now $\|\overline{\chi} - \chi_{|\overline{D}}\|_{C^1} < \min\{\tilde{\gamma}_1, d/2\}$ then condition (i) is also satisfied for $\overline{\chi}$, with $d/2$ instead of d. □

In the proof of Theorem 2.4 we made use of the following two technical remarks.

2.6. Remark. Let (M, d) be a metric space, and let $(\varphi_k)_{k \in \mathbb{N}_0}$ be a family of functions $M \to \mathbb{R}$. Assume that the φ_k are uniformly bounded and uniformly Lipschitz with Lipschitz constant $L > 0$. Then the function $\varphi : M \to \mathbb{R}$ defined by
$$\varphi(x) := \limsup_{k \to \infty} \varphi_k(x)$$
also has Lipschitz constant L.

Proof. Let $x, y \in M$. We have to show $|\varphi(x) - \varphi(y)| \leq L d(x, y)$. Assume without loss of generality that $\varphi(x) > \varphi(y)$. There exists a subsequence (φ_{k_l}) of (φ_k) with $\varphi_{k_l}(x) \to \varphi(x)$ as $l \to \infty$. Note that
$$\varphi(y) = \limsup_{k \to \infty} \varphi_k(y) \geq \limsup_{l \to \infty} \varphi_{k_l}(y).$$
For $l \in \mathbb{N}$, we have $\varphi_{k_l}(x) = \varphi_{k_l}(x) - \varphi_{k_l}(y) + \varphi_{k_l}(y) \leq L \cdot d(x,y) + \varphi_{k_l}(y)$, and thus
$$\varphi(x) = \lim_{l \to \infty} \varphi_{k_l}(x) = \limsup_{l \to \infty} \varphi_{k_l}(x) \leq L \cdot d(x, y) + \limsup_{l \to \infty} \varphi_{k_l}(y)$$
$$\leq L d(x, y) + \varphi(y),$$
and hence $|\varphi(x) - \varphi(y)| = \varphi(x) - \varphi(y) \leq L d(x, y)$. $\square$

2.7. Remark. Let (M, d) be a metric space and let $(\varphi_k)_{k \in \mathbb{N}_0}$ be a uniformly bounded sequence of functions $M \to \mathbb{R}$. Let $(D_k)_{k \in \mathbb{N}_0}$ be a sequence of closed subsets of $M \times \mathbb{R}$ such that $D_{k+1} \subset D_k$ ($k \in \mathbb{N}_0$), and
$$\forall k \in \mathbb{N}_0 : \quad \text{graph } \varphi_k \subset D_k.$$
Then $\varphi := \limsup \varphi_k$ satisfies $\text{graph } \varphi \subset \bigcap_{k=0}^{\infty} D_k$.

Proof. Let $x \in M$. There exists a subsequence $(\varphi_{k_l})_{l \in \mathbb{N}_0}$ of (φ_k) with $\varphi_{k_l}(x) \to \varphi(x)$ as $l \to \infty$. Let $n \in \mathbb{N}$. For $l \geq n$ one has $k_l \geq n$, and $\text{graph } \varphi_{k_l} \subset D_{k_l} \subset D_n$. Thus we have
$$\forall l \geq n : (x, \varphi_{k_l}(x)) \in D_n.$$
Since D_n is closed, we obtain $(x, \varphi(x)) = \lim_{l \to \infty} (x, \varphi_{k_l}(x)) \in D_n$. Since $n \in \mathbb{N}$ was arbitrary, we have $(x, \varphi(x)) \in \bigcap_{n=0}^{\infty} D_n$. $\square$

SECTION 3. COMPOSITION OF 'LOCAL' AND 'GLOBAL' MAPS

In this section we consider maps of the form $\mathcal{T}_1 \circ \mathcal{T}_0$, where the 'local' map $\mathcal{T}_0$ is given by a semigroup of linear operators and a time map τ, and the 'global' map $\mathcal{T}_1$ is nonlinear.

Let $(\tilde{S}, |\ |)$ be a normed space over $\mathbb{R}$. Consider the vector space $X = \tilde{S} \times \mathbb{C} \times \mathbb{R}$ over $\mathbb{R}$. We use the notation $(\tilde{x}, w, z)$ for elements of X, with $\tilde{x} \in \tilde{S}$, $w \in \mathbb{C}$, $z \in \mathbb{R}$. The norm on X is given by

$$|(\tilde{x}, w, z)| := \max\{|\tilde{x}|, |w|, |z|\}.$$

Let $T : \mathbb{R}_0^+ \to L_c(X, X)$ be a semigroup of linear operators (i.e., $T(t+s) = T(t)T(s)$, $t, s \geq 0$) satisfying the subsequent conditions:

(T1) The decomposition $X = \tilde{S} \times \mathbb{C} \times \mathbb{R}$ is invariant under $T(t)$ for all $t \geq 0$. (We write $\tilde{T}(t)$ for the restriction of $T(t)$ to $\tilde{S}$.)

(T2) There exist $\lambda > 0$, $\mu = \rho + i\omega \in \mathbb{C}$ and $\gamma < 0, K > 0$ with the following properties:

$$\omega \neq 0, \quad \gamma < \rho < 0,$$

and for $t \geq 0$ and $(\tilde{x}, w, z) \in X$ one has

$$T(t)(\tilde{x}, w, z) = (\tilde{y}, \exp(\mu t) \cdot w, \exp(\lambda t) \cdot z),$$

where $\tilde{y} = \tilde{T}(t)\tilde{x}$, and

$$|\tilde{y}| \leq K \cdot \exp(\gamma t)|\tilde{x}|.$$

(T3) There exist $t_0 > 0$ and $\tilde{K} > 0$ such that the map

$$[t_0, \infty) \times \tilde{S} \ni (s, \tilde{x}) \mapsto \tilde{T}(s)\tilde{x} \in \tilde{S}$$

is continuously differentiable, and for $t \geq t_0, \tilde{x} \in \tilde{S}$ one has

$$|\frac{d}{ds}(s \mapsto \tilde{T}(s)\tilde{x})|_{s=t}| \leq \tilde{K} \cdot \exp(\gamma t) \cdot |\tilde{x}|.$$

(T4) $|\rho| < \lambda$.

Comment: In the application of the result from this section, conditions (T1) and (T2) correspond to the following shape of the spectrum of the infinitesimal generator A of T: One positive real eigenvalue λ, and a sequence of conjugate complex pairs $(\mu_k, \overline{\mu}_k)_{k=1}^\infty$ of eigenvalues with negative real part, decreasing with increasing k. The space $\mathbb{C}$ in the decomposition of X corresponds to the 'leading pair' of stable eigenvalues, $(\mu_1, \overline{\mu}_1)$. Condition

(T4) expresses that the expansion in the unstable direction is stronger than the contraction in the stable direction.

Let $z_0 > 0$ be given. Define $\tau(z) := \dfrac{1}{\lambda} \log \dfrac{z_0}{|z|}$ for $z \in [-z_0, z_0] \setminus \{0\}$, and define $\mathcal{T}_0 : \tilde{S} \times \mathbb{C} \times ([-z_0, z_0] \setminus \{0\}) \to \tilde{S} \times \mathbb{C} \times \{z_0, -z_0\}$ by

$$\mathcal{T}_0(\tilde{x}, w, z) := T(\tau(z))(\tilde{x}, w, z).$$

Then

$$\mathcal{T}_0(\tilde{x}, w, z) = (\tilde{T}(\tau(z))\tilde{x}, \exp(\mu\tau(z)) \cdot w, \exp(\lambda \cdot \tau(z)) \cdot z)$$
$$= (\tilde{T}(\tau(z))\tilde{x}, \exp(\mu\tau(z)) \cdot w, \operatorname{sign}(z) \cdot z_0).$$

$\mathcal{T}_0$ is our 'local' map, given by the semigroup T and the time until the third component reaches z_0 in absolute value. We now turn to the definition of the 'global' map, and to the composition of both.

For $r > 0$, set $\mathbb{C}_r := \{\zeta \in \mathbb{C} \mid |\zeta| = r\}$. Let Σ be an open neighborhood of $(0,0)$ in $\tilde{S} \times \mathbb{C}$. Let $R > 0$ and let $G^+, G^- : \Sigma \to \tilde{S} \times \mathbb{C}_R \times \mathbb{R} \subset X$ be BC^1 maps with the following properties. (The notation '$\pm$' means that a condition is to be read as two conditions, one for the case '+' and one for the case '−'.)

(G1) $G^{\pm}(0,0) \in \tilde{S} \times \mathbb{C}_R \times \{0\}$.

(G2) The linear map $DG^{\pm}(0,0) \in L_c(\tilde{S} \times \mathbb{C}, \tilde{S} \times \mathbb{C} \times \mathbb{R})$ satisfies

$$DG^{\pm}(0,0)(\{0\} \times \mathbb{C}) \not\subset \tilde{S} \times \mathbb{C} \times \{0\}.$$

Condition (G1) implies the existence of unique $\varphi_-, \varphi_+ \in [0, 2\pi)$ and $\tilde{x}_+, \tilde{x}_- \in \tilde{S}$ such that

$$G^{\pm}(0,0) = (\tilde{x}_{\pm}, R \cdot \exp(i\varphi_{\pm}), 0).$$

Define the 'global' map $\mathcal{T}_1 : \Sigma \times \{z_0, -z_0\} \to \tilde{S} \times \mathbb{C}_R \times \mathbb{R}$ by

$$\mathcal{T}_1(\tilde{y}, w, z_0) := G^+(\tilde{y}, w),$$
$$\mathcal{T}_1(\tilde{y}, w, -z_0) := G^-(\tilde{y}, w) \text{ for } (\tilde{y}, w) \in \Sigma.$$

3.1. Remark. *a) The first and second components of $\mathcal{T}_0$ are C^1 functions on the set*

$$\mathcal{D}_0 := \{(\tilde{x}, w, z) \in \tilde{S} \times \mathbb{C} \times ([-z_0, z_0] \setminus \{0\}) \mid \tau(z) > t_0\}.$$

b) If $M \subset \mathcal{D}_0$ is a bounded subset with $\operatorname{dist}(\operatorname{pr}_3 M, \{0\}) > 0$ (where pr_3 denotes the projection onto the third factor in $\tilde{S} \times \mathbb{C} \times \mathbb{R}$), then the derivatives of these two components of $\mathcal{T}_0$ are bounded on M.

Proof. Ad a): The second component of $\mathcal{T}_0$ is C^1 on all of its domain, since τ is C^1. On the set $\mathcal{D}_0$, the first component of $\mathcal{T}_0$ is given by the composition

of the C^1 map $\mathcal{D}_0 \ni (\tilde{x}, w, z) \mapsto (\tau(z), \tilde{x}) \in (t_0, \infty) \times \tilde{S}$ with the map from assumption (T3) and is thus C^1.

Ad b): There exists $\zeta > 0$ with $|z| \geq \zeta$ for all $(\tilde{x}, w, z) \in M$. Since τ and the derivative τ' are bounded on the set $\{z \in [-z_0, z_0] \mid |z| \geq \zeta\}$, the assertion follows from the formula for $\mathcal{T}_0$, from boundedness of M, and from the estimates in conditions (T2) and (T3). □

Next we show that the composition of the maps $\mathcal{T}_1$ and $\mathcal{T}_0$ leads to a map that satisfies the conditions of Theorem 2.4 (and thus has symbolic dynamics on appropriate subsets).

3.2. Lemma. *Assume the above conditions* (T1)–(T4), *let* $z_0 > 0$, *the maps* $G^\pm$, *and* $R > 0$ *and* $\varphi_\pm \in [0, 2\pi)$ *be as above, and assume that* (G1) *and* (G2) *hold. Let* $\tilde{r} > 0$, $\tilde{\delta} \in (0, \pi/2)$ *be given.*

There exist numbers $\delta \in (0, \tilde{\delta})$, $k_0 \in \mathbb{N}$, *an open set* $D \subset \tilde{S} \times \mathbb{R} \times (-z_0, z_0) \setminus \{0\}$ *and intervals*

$$J_{l,+}^{(k)} = [a_{l,+}^{(k)}, b_{l,+}^{(k)}] \subset \mathbb{R}^+, \quad J_{l,-}^{(k)} = [b_{l,-}^{(k)}, a_{l,-}^{(k)}] \subset \mathbb{R}^- \quad (k \geq k_0,\, l = +, -)$$

such that the following statements are true.

A1) *For all* $k \geq k_0$ *and* $l, m = +, -$, *one has*

$$J_{l,m}^{(k+1)} = \exp(-2\pi\lambda/\omega) \cdot J_{l,m}^{(k)}.$$

Further, with

$$I_l := \{\tilde{x} \in \tilde{S} \mid |\tilde{x} - \tilde{x}_l| \leq \tilde{r}\} \times [\varphi_l - \delta, \varphi_l + \delta] \quad (l = +, -), \text{ and}$$

$$\Delta^{(k)} := \bigcup_{l,m=+,-} I_l \times J_{l,m}^{(k)} \text{ for } k \geq k_0,$$

the set D *is an open neighborhood of* $\Delta^{(k)}$ *for all* $k \geq k_0$.

A2) *For* $(\tilde{x}, \theta, z) \in D$, *one has*

$$\mathcal{T}_0(\tilde{x}, R \cdot \exp(i\theta), z) \in \Sigma \times \{-z_0, z_0\}.$$

A3) *The map* $\kappa : D \cup [(I_- \cup I_+) \times \mathbb{R}] \to \tilde{S} \times \mathbb{C} \times \mathbb{R}$,

$$\kappa(\tilde{x}, \theta, z) := (\tilde{x}, R \cdot \exp(i\theta), z)$$

is injective, and $(\mathcal{T}_1 \circ \mathcal{T}_0 \circ \kappa)(D) \subset \kappa((I_- \cup I_+) \times \mathbb{R})$. *The inverse map* $\kappa^{-1} : \text{image}(\kappa) \to D \cup [(I_- \cup I_+) \times \mathbb{R}] \subset \tilde{S} \times \mathbb{R} \times \mathbb{R}$ *is Lipschitz continuous.*

A4) *The map* $\chi : D \to \tilde{S} \times \mathbb{R} \times \mathbb{R}$,

$$\chi(\tilde{x}, \theta, z) := (\kappa^{-1} \circ \mathcal{T}_1 \circ \mathcal{T}_0 \circ \kappa)(\tilde{x}, \theta, z)$$

if of class C^1 *and satisfies the assumptions of Theorem 2.4 on each set* $\Delta^{(k)}$, $k \geq k_0$.

There exists $\tilde{L} > 0$ *such that, on each* $\Delta^{(k)}$, *the constant* $L = L^{(k)}$ *from condition* (iii) *of Theorem 2.4 can be chosen as* $L^{(k)} = \tilde{L} \cdot \exp(-2\pi\lambda k/\omega)$.

Proof. We write pr_j for the projection onto the j-th factor in $\tilde{S}\times\mathbb{C}$, $j=1,2$, and in $\tilde{S}\times\mathbb{C}\times\mathbb{R}$ and $\tilde{S}\times\mathbb{R}\times\mathbb{R}$, $j=1,2,3$. As before, statements including the symbol '$\pm$' are to be read as two statements, one for the case '+' and one for the case '−'.

1. Condition (G2) shows that the linear operator $\mathrm{pr}_3 \circ DG^{\pm}(0,0)_{|\{0\}\times\mathbb{C}}$ is nonzero, and therefore surjective. The domain of this operator is two-dimensional (over $\mathbb{R}$) and its range is one-dimensional, so the kernel has dimension one. It follows that there exist unique $\theta_{0,+}$ and $\theta_{0,-} \in [0,\pi)$ with

(3.2.1) $\quad \mathrm{pr}_3 DG^+(0,0)(0, \exp(i\theta_{0,+})) = 0 = \mathrm{pr}_3 DG^-(0,0)(0, \exp(i\theta_{0,-}))$.

Further, for $\theta \in [0, 2\pi)$, we have the equivalence

(3.2.2) $\quad \mathrm{pr}_3 DG^{\pm}(0,0)(0, \exp(i\theta)) = 0 \iff \theta \in \{\theta_{0,\pm}, \theta_{0,\pm}+\pi\}$.

In particular, we have

$$\frac{d}{d\theta}[\mathrm{pr}_3 DG^{\pm}(0,0)(0, \exp(i\theta))]|_{\theta=\theta_{0,\pm}}$$
$$= \mathrm{pr}_3 DG^{\pm}(0,0)(0, i\exp(i\theta_{0,\pm}))$$
$$= \mathrm{pr}_3 DG^{\pm}(0,0)[0, \exp(i(\theta_{0,\pm}+\pi/2))]$$
$$\neq 0.$$

It follows that there exist $c_0 > 0$ and $\delta_0 \in (0, \tilde{\delta})$ such that one has

(3.2.3) $\quad \forall \theta \in (-\delta_0, \delta_0): \quad |\mathrm{pr}_3 DG^{\pm}(0,0)[0, \exp(i(\theta_{0,\pm}+\theta))]| \geq c_0|\theta|$.

2. Since the characteristic value μ from condition (T2) is not real, there exists precisely one $\nu \in [0, 2\pi) \setminus \{0, \pi\}$ such that

$$\mu = |\mu| \cdot \exp(i\nu).$$

Set $G_0 := \max\{|DG^+(0,0)|, |DG^-(0,0)|\}$. Since $\nu \notin \pi \cdot \mathbb{Z}$, we see from (3.2.2) that
$$\mathrm{pr}_3 DG^{\pm}(0,0)[0, \exp(i(\theta_{0,\pm}+\nu))] \neq 0.$$

It follows that there exist $\delta_1 \in (0, \delta_0]$ and $c_1 > 0$ such that the subsequent statements (3.2.4) and (3.2.5) hold.
(3.2.4)
  a) $\{(\tilde{y}, w) \in \tilde{S} \times \mathbb{C} \mid |\tilde{y}| \leq \delta_1, |w| \leq \delta_1\} \subset \Sigma$.
  b) $\exp(i(\varphi_+ + [-\delta_1, \delta_1])) \cap \exp(i(\varphi_- + [-\delta_1, \delta_1])) = \emptyset$ in case $\varphi_+ \neq \varphi_-$.
  c) $\dfrac{c_0 \delta_1}{4G_0} \leq 1$.

(3.2.5) $\forall \tilde{x}, \tilde{y} \in \tilde{S}, |\tilde{x}|, |\tilde{y}| \leq \delta_1 \ \forall w \in \mathbb{C}, |w| \leq \delta_1 \ \forall \theta \in \mathbb{R}, |\theta| \leq \delta_1 :$
$$|\mathrm{pr}_3 DG^\pm(\tilde{x}, w)[\tilde{y}, \exp(i(\theta_{0,\pm} + \nu + \theta))]| \geq c_1 > 0.$$

Set $\pi^\pm(\theta) := \mathrm{pr}_3 DG^\pm(0,0)[0, \exp(i(\theta_{0,\pm} + \theta))]$ for $\theta \in [-\delta_1, \delta_1]$. It follows then from (3.2.2), (3.2.3) and from $\delta_1 \leq \delta_0$ that

(3.2.6) $\quad \mathrm{sign}\,(\pi^\pm(\theta)) = -\mathrm{sign}\,\pi^\pm(-\theta)$ for $\theta \in (0, \delta_1]$, and
$$|\pi^\pm(\theta)| \geq c_0 |\theta| \text{ for } \theta \in [-\delta_1, \delta_1].$$

3. There exist neighborhoods $N^\pm$ of $(R, \varphi_\pm)$ in $\mathbb{R}^2$ such that the maps

$$q^\pm : N^\pm \to \mathbb{C}, \quad (r, \varphi) \mapsto r \cdot \exp(i\varphi)$$

are Lipschitz continuous diffeomorphisms onto their images $Q^\pm := q^\pm(N^\pm)$, with Lipschitz continuous inverse diffeomorphisms and with uniformly continuous derivatives. There exist BC^1 functions

$$\arg^\pm : Q^\pm \to \mathbb{R}$$

such that the inverse diffeomorphisms are given by $Q^\pm \ni z \mapsto (|z|, \arg^\pm(z))$. We can choose $r_0 \in (0, \delta_1]$ and $d_1 > 0$ such that for $(\tilde{y}, w) \in \tilde{S} \times \mathbb{C}$ with $|\tilde{y}| \leq r_0$, $|w| \leq r_0$, the subsequent three properties hold. (Recall the given number $\tilde{r} > 0$.)

(3.2.7) $\qquad\qquad\qquad |DG^\pm(\tilde{y}, w)| \leq 2G_0.$

If we set $(\hat{x}, \hat{w}, \hat{z}) := G^\pm(\tilde{y}, w)$ then

(3.2.8) $\quad |\hat{x} - \tilde{x}_\pm| \leq \dfrac{\tilde{r}}{2},$ and $\hat{w} + \mathbb{C}(d_1) \subset Q^\pm$, $|\hat{w}| = R$, and
$$|\arg^\pm(\hat{w}) - \varphi_\pm| \leq \delta_1/8.$$

(3.2.9) $|G^\pm(\tilde{y}, w) - (\tilde{x}_\pm, R \cdot \exp(i\varphi_\pm), 0) - DG^\pm(0,0)(\tilde{y}, w)| \leq \dfrac{c_0 \delta_1}{8} |(\tilde{y}, w)|.$

(We now prove an analog of property (3.2.6) for the nonlinear maps $G^\pm$.)
(3.2.10) *Claim:* For $r \in (0, r_0]$, for $\theta \in [\delta_1/2, \delta_1] \cup [-\delta_1, -\delta_1/2]$ and for $\tilde{y} \in \tilde{S}$ with $|\tilde{y}| \leq \dfrac{c_0 \delta_1}{8G_0} \cdot r$, one has

$$|\mathrm{pr}_3 G^\pm(\tilde{y}, r \cdot \exp(i(\theta_{0,\pm} + \theta)))| \geq \frac{c_0 \delta_1}{4} \cdot r, \text{ and}$$
$$\mathrm{sign}\,\mathrm{pr}_3 G^\pm(\tilde{y}, r \cdot \exp(i(\theta_{0,\pm} + \theta))) = \mathrm{sign}\,\pi^\pm(\theta).$$

Proof: Let $r, \theta, \tilde{y}$ be as in the claim. Then (3.2.4)c) implies $|\tilde{y}| \leq r$. In view of condition (G1), we have

$$\text{pr}_3 G^{\pm}(\tilde{y}, r \cdot \exp(i(\theta_{0,\pm} + \theta)))$$
$$= \text{pr}_3[G^{\pm}(\tilde{y}, r \cdot \exp(i(\theta_{0,\pm} + \theta))) - G^{\pm}(0,0)]$$
$$= \text{pr}_3 DG^{\pm}(0,0)[\tilde{y}, r \cdot \exp(i(\theta_{0,\pm} + \theta))] + \rho_1^{\pm}(\tilde{y}, r, \theta),$$

where the properties (3.2.9), $|\text{pr}_3| \leq 1$, and $|\tilde{y}| \leq r$ imply that the remainder term satisfies

$$|\rho_1^{\pm}(\tilde{y}, r, \theta)| \leq \frac{c_0 \delta_1}{8} |(\tilde{y}, r \cdot \exp(i(\theta_{0,\pm} + \theta)))| = \frac{c_0 \delta_1}{8} \cdot r.$$

Further,

$$\text{pr}_3 DG^{\pm}(0,0)(\tilde{y}, r \cdot \exp(i(\theta_{0,\pm} + \theta)))$$
$$= \text{pr}_3 DG^{\pm}(0,0)(0, r \cdot \exp(i(\theta_{0,\pm} + \theta))) + \text{pr}_3 DG^{\pm}(0,0)(\tilde{y}, 0)$$
$$= r \cdot \pi^{\pm}(\theta) + \text{pr}_3 DG^{\pm}(0,0)(\tilde{y}, 0),$$

and the last term satisfies

$$|\text{pr}_3 DG^{\pm}(0,0)(\tilde{y}, 0)| \leq G_0 \cdot |\tilde{y}| \leq \frac{c_0 \delta_1}{8} \cdot r.$$

Together, we obtain from the above estimates that

$$\text{pr}_3 G^{\pm}(\tilde{y}, r \cdot \exp(i(\theta_{0,\pm} + \theta))) = r \cdot \pi^{\pm}(\theta) + \rho^{\pm}(\tilde{y}, r, \theta),$$

where the second term satisfies $|\rho^{\pm}(\tilde{y}, r, \theta)| \leq \frac{c_0 \delta_1}{4} \cdot r$. It follows from (3.2.6) and from $\delta_1 \geq |\theta| \geq \delta_1/2$ that

$$\text{sign } \text{pr}_3 G^{\pm}(\tilde{y}, r \cdot \exp(i(\theta_{0,\pm} + \theta))) = \text{sign } \pi^{\pm}(\theta),$$

and $|\text{pr}_3 G^{\pm}(\tilde{y}, r \cdot \exp(i(\theta_{0,\pm} + \theta)))| \geq c_0 \cdot r \cdot |\theta| - \frac{c_0 \delta_1}{4} \cdot r$
$$\geq \frac{c_0 \delta_1}{4} \cdot r.$$

(Claim (3.2.10) is proved.)

4. Recall the numbers $\lambda > 0$, $\mu = \rho + i\omega$, and the properties $0 < -\rho < \lambda, \gamma < \rho$, and the time map τ. The number δ from the assertion of the lemma is defined by

$$\delta := \delta_1/4.$$

Set $\tilde{R} := \max\{|\tilde{x}_+|, |\tilde{x}_-|\}$, and set $A := \max\{\|D \arg^+\|_\infty, \|D \arg^-\|_\infty\}$. Choose $k_0 \in \mathbb{N}, k_0 \geq 2$ such that for $k \in \mathbb{N}, k \geq k_0$, one has

$2 \cdot \exp(-\frac{\lambda}{\omega}\pi(2k-3)) < 1$, and that for all $z \in \mathbb{R}$ with $|z| \in (0, z_0 \cdot 2 \cdot \exp(-\frac{\lambda}{\omega}\pi(2k-3))]$ the following estimates hold.
(3.2.11)
  a) $K \cdot \exp(\gamma\tau(z)) \max\{1, \tilde{R}+2\tilde{r}\} \leq R \cdot \exp(\rho\tau(z)) \leq r_0/2$.
  b) $K \cdot \exp(\gamma\tau(z))(\tilde{R}+\tilde{r}) \leq \frac{c_0\delta_1}{8G_0} \cdot R \cdot \exp(\rho\tau(z))$.
  c) $R \cdot \frac{c_0\delta_1}{4} \exp(\frac{\rho}{\omega}(2k+3)\pi) > z_0 \cdot \exp(-\frac{\lambda}{\omega}(2k-3)\pi)$.
  d) With t_0 from (T3), one has $\tau(z) > t_0$.
  e) With $\tilde{K}$ from (T3), one has $\tilde{K}(\tilde{R}+\tilde{r})\exp(\gamma\tau(z)) \leq R\cdot\exp(\rho\tau(z))\cdot|\mu|$.
  f) $\frac{\tilde{K}(\tilde{R}+\tilde{r})}{R|\mu|}\exp[(\gamma-\rho)\tau(z)] \leq \delta_1$.
  g) $|z|(1+A) \cdot \frac{16G_0 \cdot \lambda}{|\mu| \cdot c_1} < \min\{1, \frac{c_1}{4G_0(1+A)}\}$.

For $k \in \mathbb{N}, k \geq k_0$, and $l = +, -$, set

$$a_{l,+}^{(k)} := z_0 \cdot \exp[-\frac{\lambda}{\omega}(2\pi k + \theta_{0,+} - \varphi_l + \frac{3\delta_1}{4})],$$

$$b_{l,+}^{(k)} := z_0 \cdot \exp[-\frac{\lambda}{\omega}(2\pi k + \theta_{0,+} - \varphi_l - \frac{3\delta_1}{4})]$$

$$a_{l,-}^{(k)} := -z_0 \cdot \exp[-\frac{\lambda}{\omega}(2\pi k + \theta_{0,-} - \varphi_l + \frac{3\delta_1}{4})]$$

$$b_{l,-}^{(k)} := -z_0 \cdot \exp[-\frac{\lambda}{\omega}(2\pi k + \theta_{0,-} - \varphi_l - \frac{3\delta_1}{4})].$$

Then, for $l = +, -$, $k \geq k_0$, we have the following properties.
(3.2.12)
  a) $\tau(a_{l,\pm}^{(k)}) = (1/\omega)(2\pi k + \theta_{0,\pm} - \varphi_l + 3\delta_1/4)$,
  b) $\tau(b_{l,\pm}^{(k)}) = (1/\omega)(2\pi k + \theta_{0,\pm} - \varphi_l - 3\delta_1/4)$.
  c) $B^{(k)} := \max_{l,m=+,-} |b_{l,m}^{(k)}| \leq z_0 \cdot \exp(-\frac{\lambda}{\omega}(2k-3)\pi)$.

With δ from above and the given number $\tilde{r} > 0$, define I_l and $J_{l,m}^{(k)}$ and $\Delta^{(k)}$ $(k \geq k_0, l, m = +, -)$ as in the assertion of the lemma. Set

$$I'_\pm := \{(\tilde{x}, \theta) \in \tilde{S} \times \mathbb{R} \mid |\tilde{x} - \tilde{x}_\pm| \leq \frac{\tilde{r}}{2}, |\theta - \varphi_\pm| \leq \frac{\delta_1}{8}\}.$$

Since $\tilde{r}/2 + \tilde{r}/2 = \tilde{r}$ and $\delta_1/8 = \delta/2$, we have with $d := \min\{\tilde{r}/2, \delta_1/8\}$ that $I'_\pm \subset \text{int}_d(I_\pm)$.
  For $l = +, -$, set

$$D_{l,+} := \{(\tilde{x}, \theta, z) \in \tilde{S} \times \mathbb{R} \times \mathbb{R} \mid |\tilde{x}| < \tilde{R}+2\tilde{r}, |\theta - \varphi_l| < 2\delta, z \in (0, 2b_{l,+}^{(k_0)})\},$$

$$D_{l,-} := \{(\tilde{x}, \theta, z) \in \tilde{S} \times \mathbb{R} \times \mathbb{R} \mid |\tilde{x}| < \tilde{R}+2\tilde{r}, |\theta - \varphi_l| < 2\delta, z \in (2b_{l,-}^{(k_0)}, 0)\}.$$

Set $D_+ := D_{-,+} \cup D_{+,+}$, and $D_- := D_{-,-} \cup D_{+,-}$, and $D := D_+ \cup D_-$. It is clear from the definitions of $\tilde{R}$ and $\Delta^{(k)}$ that D is an open neighborhood of $\Delta^{(k)}$ in $\tilde{S} \times \mathbb{R} \times \mathbb{R}$, for all $k \geq k_0$.

5. Assertion A1) is obviously true. Fix now $k \geq k_0$. We omit the superscript (k) from now on, and we prove assertions A2)–A4).

Proof of A2): Let $(\tilde{x}, \theta, z) \in D$. It follows from (3.2.12)c) and (3.2.11)d) that $\tau(z) > t_0 > 0$. Further, with $\tilde{y} := \tilde{T}(\tau(z))\tilde{x}$, we have

$$T(\tau(z))(\tilde{x}, R \cdot \exp(i\theta), z) = (\tilde{y}, R \cdot \exp(\rho\tau(z)) \cdot \exp(i(\omega\tau(z) + \theta)), \operatorname{sign}(z)z_0).$$

From (T2) and (3.2.11)a) and the definition of D, we obtain

$$(3.2.13) \quad |\tilde{y}| \leq K \cdot \exp(\gamma\tau(z))|\tilde{x}| \leq K \cdot \exp(\gamma\tau(z))(\tilde{R} + 2\tilde{r}) \leq \frac{r_0}{2} \leq r_0, \text{ and}$$

$$(3.2.14) \qquad R \cdot \exp(\rho\tau(z)) \leq \frac{r_0}{2} \leq r_0.$$

It follows from (3.2.4)a) and from $r_0 \leq \delta_1$ that

$$T(\tau(z))(\tilde{x}, R \cdot \exp(i\theta), z) \in \Sigma \times \{z_0, -z_0\}.$$

Proof of A3): For $(\tilde{x}, \theta, z) \in D \cup [(I_- \cup I_+) \times \mathbb{R}]$, we have

$$\theta \in \{\varphi_+, \varphi_-\} + (-2\delta, 2\delta) = \{\varphi_+, \varphi_-\} + (-\frac{\delta_1}{2}, \frac{\delta_1}{2}),$$

and $\kappa(\tilde{x}, \theta, z) = (\tilde{x}, R \cdot \exp(i\theta), z)$. Since $0 < \delta_1 \leq \delta_0 \leq \pi/2$, the map $\theta \mapsto R \cdot \exp(i\theta)$ is injective on both intervals $\varphi_+ + (-\delta_1/2, \delta_1/2)$ and $\varphi_- + (-\delta_1/2, \delta_1/2)$. One sees now from property (3.2.4)b), and from $\delta_1 \leq \pi/2$, that κ is injective and that κ^{-1} is Lipschitz continuous. Assume now that $(\tilde{x}, \theta, z) \in D_+$. Set

$$\hat{x} := \operatorname{pr}_1(\mathcal{T}_1 \circ \mathcal{T}_0 \circ \kappa)(\tilde{x}, \theta, z), \quad \hat{z} := \operatorname{pr}_3(\mathcal{T}_1 \circ \mathcal{T}_0 \circ \kappa)(\tilde{x}, \theta, z).$$

Define $(\tilde{y}, w) \in \tilde{S} \times \mathbb{C}$ by

$$(\tilde{y}, w, z_0) = \mathcal{T}_0(\tilde{x}, R \cdot \exp(i\theta), z) = (\mathcal{T}_0 \circ \kappa)(\tilde{x}, \theta, z).$$

Using (3.2.13), (3.2.14) and property (3.2.8) of r_0, we see that $G^+(\tilde{y}, w) = (\hat{x}, \hat{w}, \hat{z})$, where $|\hat{w}| = R$, $\hat{w} + \mathbb{C}(d_1) \subset Q^+$, and $|\hat{x} - \tilde{x}_+| \leq \frac{\tilde{r}}{2}$. Further, with

$$(3.2.15) \qquad \hat{\theta} := \arg^+(\hat{w}),$$

we have $\hat{w} = R \cdot \exp(i\hat{\theta})$ and $\hat{\theta} \in [\varphi_+ - \delta_1/8, \varphi_+ + \delta_1/8]$. Consequently, $(\hat{x}, \hat{\theta}) \in I'_+$. In particular, one has

$$(\mathcal{T}_1 \circ \mathcal{T}_0 \circ \kappa)(\tilde{x}, \theta, z) = \kappa(\hat{x}, \hat{\theta}, \hat{z}) \in \kappa(I_+ \times \mathbb{R}).$$

In case $(\tilde{x}, \theta, z) \in D_-$, the proof that the corresponding $(\hat{x}, \hat{\theta})$ satisfies $(\hat{x}, \hat{\theta}) \in I'_-$, and that $(\mathcal{T}_1 \circ \mathcal{T}_0 \circ \kappa)(\tilde{x}, \theta, z) \in \kappa(I_- \times \mathbb{R})$ is analogous. One has to replace '+' by '−' and z_0 by $-z_0$.

Proof of A4): First, note that the indices $+, -$ of I_+, I_- and $J_{l,m}^{(k)}$ ($l, m = +, -$) correspond to the indices $1, -1$ in Theorem 2.4. The choice of δ_1 and of $\delta = \delta_1/4$ implies that $\text{dist}(I_+, I_-) > 0$ if $\varphi_+ \neq \varphi_-$, and that $I_+ = I_-$ if $\varphi_+ = \varphi_-$, as required for Theorem 2.4.

Next, we show that the map χ is C^1. For $(\tilde{x}, \theta, z) \in D_+$ we have

$$(\mathcal{T}_1 \circ \mathcal{T}_0 \circ \kappa)(\tilde{x}, \theta, z) = G^+(\text{pr}_1 \mathcal{T}_0(\tilde{x}, R \cdot \exp(i\theta), z), \text{pr}_2 \mathcal{T}_0(...)).$$

Note that (3.2.12)c) and the choice of k_0 imply $2B < z_0$. Hence, the definition of D and property (3.2.11)d) imply that with $\mathcal{D}_0$ from Remark 3.1, one has $(\tilde{x}, R \cdot \exp(i\theta), z) \in \mathcal{D}_0$. Since G^+ is C^1, we obtain from Remark 3.1 that $\mathcal{T}_1 \circ \mathcal{T}_0 \circ \kappa$ is C^1 on D_+. In view of (3.2.15), we have on D_+

$$\chi(\tilde{x}, \theta, z)$$
$$= (\text{pr}_1(\mathcal{T}_1 \circ \mathcal{T}_0 \circ \kappa)(\tilde{x}, \theta, z), \text{arg}^+(\text{pr}_2(\mathcal{T}_1 \circ \mathcal{T}_0 v \circ \kappa)(...), \text{pr}_3(\mathcal{T}_1 \circ \mathcal{T}_0 \circ \kappa)(...)),$$

which shows that χ is C^1 on D_+. Similarly, χ is C^1 on D_-. We now check conditions (i)-(iii) of Theorem 2.4. Note that S in that theorem corresponds to $\tilde{S} \times \mathbb{R}$ now, and pr_2 in that theorem corresponds to pr_3 in the present situation. In the proof of A3) we have already shown that for $(\tilde{x}, \theta, z) \in D$ and $\hat{x} := \text{pr}_1 \chi(\tilde{x}, \theta, z)$, $\hat{\theta} := \text{pr}_2 \chi(\tilde{x}, \theta, z)$, one has $(\hat{x}, \hat{\theta}) \in I'_+ \cup I'_-$, and we know that $I'_l \subset \text{int}_d(I_l)$, $l = +, -$. Thus condition (i) holds even for all $(\tilde{x}, \theta, z) \in D$.

Ad condition (ii): Let $(\tilde{x}, \theta, z) \in \Delta$, assume first that $z > 0$, and define $(\tilde{y}, w) \in \Sigma$ by $(\tilde{y}, w, z_0) = \mathcal{T}_0(\tilde{x}, R \cdot \exp(i\theta), z)$. As in the proof of A2) (see (3.2.13)), we have with

$$r(z) := R \cdot \exp(\rho \tau(z))$$

that $r(z) \leq r_0$, and $w = r(z) \cdot \exp[i(\omega \tau(z) + \theta)]$. The definition of Δ implies $|\tilde{x}| \leq \tilde{R} + \tilde{r}$. From (3.2.13) and (3.2.11)b), one obtains

$$(3.2.16) \qquad |\tilde{y}| \leq K \cdot \exp(\gamma \tau(z))(\tilde{R} + \tilde{r}) \leq \frac{c_0 \delta_1}{8 G_0} \cdot r(z).$$

In the verification of condition (ii), we have to distinguish two cases.

First case: $z = b_{l,+}$ for an $l \in \{-, +\}$. The definition of Δ implies $|\theta - \varphi_l| \leq \delta$. In view of (3.2.12)b), one has

$$\omega \tau(z) + \theta = 2\pi k + \theta_{0,+} - \varphi_l - 3\delta_1/4 + \theta.$$

Setting $\beta_l(\theta) := \theta_{0,+} - \varphi_l - 3\delta_1/4 + \theta$, we know

(3.2.17) $$w = r(z) \cdot \exp(i\beta_l(\theta)), \text{ and}$$

(3.2.18) $$\beta_l(\theta) \in \theta_{0,+} + [-\delta, \delta] - 3\delta_1/4 \subset \theta_{0,+} + [-\delta_1, -\delta_1/2].$$

Combining (3.2.16) and (3.2.18) with Claim (3.2.10), we infer that

$$|\mathrm{pr}_3 G^+(\tilde{y}, w)| \geq \frac{c_0 \delta_1}{4} \cdot r(b_{l,+}).$$

Second case: $z = a_{l,+}$ for an $l \in \{-, +\}$. Then one obtains similarly, using (3.2.12)a), that $\omega\tau(z) + \theta = 2\pi k + \alpha_l(\theta)$, where $\alpha_l(\theta)$ satisfies

(3.2.19) $$\alpha_l(\theta) \in \theta_{0,+} + [\delta_1/2, \delta_1], \text{ and } w = r(z) \cdot \exp(i\alpha_l(\theta)).$$

Further, using Claim (3.2.10), one gets

$$|\mathrm{pr}_3 G^+(\tilde{y}, w)| \geq \frac{c_0 \delta_1}{4} r(a_{l,+}).$$

Note now that for $l \in \{-, +\}$ and for $c \in \{a_{l,+}, b_{l,+}\}$, it follows from $|\theta_{0,+} - \varphi_l| \leq 2\pi$, from $3\delta_1/4 \leq \pi$ and from (3.2.12)a), b), that

$$r(c) = R \cdot \exp(\rho\tau(c)) \geq R \cdot \exp(\frac{\rho}{\omega}(2k+3)\pi).$$

Using (3.2.11)c) and (3.2.12)c), we obtain that in both cases one has

$$|\mathrm{pr}_3 \chi(\tilde{x}, \theta, z)| = |\mathrm{pr}_3 G^+(\tilde{y}, w)| \geq \frac{c_0 \delta_1}{4} R \cdot \exp(\frac{\rho}{\omega}(2k+3)\pi)$$
$$> z_0 \cdot \exp(-\frac{\lambda}{\omega}(2k-3)\pi) \geq B.$$

Hence, the inequalities from condition (ii) of Theorem 2.4 hold in case $z > 0$. We now check the sign conditions: Combining (3.2.16) and the second assertion of Claim (3.2.10) with (3.2.17)–(3.2.19), we see that for $l = +, -$

$$\mathrm{sign}\, \mathrm{pr}_3\chi(\tilde{x}, \theta, a_{l,+}) = \mathrm{sign}\, \pi^+(\alpha_l(\theta) - \theta_{0,+}),$$
$$\mathrm{sign}\, \mathrm{pr}_3\chi(\tilde{x}, \theta, b_{l,+}) = \mathrm{sign}\, \pi^+(\beta_l(\theta) - \theta_{0,+}).$$

It follows from the first line of (3.2.6), together with (3.2.18), (3.2.19), that

$$\mathrm{sign}\, \mathrm{pr}_3\chi(\tilde{x}, \theta, a_{l,+}) = -\mathrm{sign}\, \mathrm{pr}_3\chi(\tilde{x}, \theta, b_{l,+}) \quad (l = -, +).$$

Condition (ii) of Theorem 2.4 is now verified for the case $z > 0$. The proof for the case $z < 0$ is analogous, replacing $a_{l,+}, b_{l,+}$ by $a_{l,-}, b_{l,-}$, replacing $\theta_{0,+}$ by $\theta_{0,-}$, and replacing G^+, π^+ by G^-, π^-.

Ad condition (iii): Let $(\tilde{x}, \theta, z) \in \Delta$. We estimate the derivatives of the components of $\mathcal{T}_0 \circ \kappa$ first: Since $\mathrm{pr}_1(\mathcal{T}_0 \circ \kappa)(\tilde{x}, \theta, z) = \tilde{T}(\tau(z))\tilde{x}$, it follows from (T2) that we have

(3.2.20) $\qquad |D_1(\mathrm{pr}_1 \circ \mathcal{T}_0 \circ \kappa)(\tilde{x}, \theta, z)| \leq K \cdot \exp(\gamma\tau(z)).$

Further,

(3.2.21) $\qquad D_2(\mathrm{pr}_1 \circ \mathcal{T}_0 \circ \kappa)(\tilde{x}, \theta, z) = 0.$

Using (T3), together with the property $\tau(z) > t_0$ (from (3.2.11)d)), and the definition of τ, we get
(3.2.22)
$$|D_3(\mathrm{pr}_1 \circ \mathcal{T}_0 \circ \kappa)(\tilde{x}, \theta, z)| \leq \tilde{K}\exp(\gamma\tau(z))|\tau'(z)||\tilde{x}| = \frac{\tilde{K}\exp(\gamma\tau(z))|\tilde{x}|}{\lambda|z|}$$
$$\leq \frac{\tilde{K}(\tilde{R}+\tilde{r})\exp(\gamma\tau(z))}{\lambda|z|}.$$

One has

(3.2.23) $\qquad D_1(\mathrm{pr}_2 \circ \mathcal{T}_0 \circ \kappa)(\tilde{x}, \theta, z) = 0,$ and

(3.2.24)
$$|D_2(\mathrm{pr}_2 \circ \mathcal{T}_0 \circ \kappa)(\tilde{x}, \theta, z)| = |\frac{d}{dt}(t \mapsto R \cdot \exp(\mu\tau(z) + it))|_{t=\theta}|$$
$$= |iR\exp(\mu\tau(z) + i\theta)|$$
$$= R \cdot \exp(\rho\tau(z)).$$

Recalling that $\mu = \rho + i\omega = |\mu| \cdot \exp(i\nu)$, we obtain the following expression for $D_3(\mathrm{pr}_2 \circ \mathcal{T}_0 \circ \kappa)$:

(3.2.25)
$$D_3(\mathrm{pr}_2 \circ \mathcal{T}_0 \circ \kappa)(\tilde{x}, \theta, z)(1)$$
$$= \mu\tau'(z)(\mathrm{pr}_2 \circ \mathcal{T}_0 \circ \kappa)(\tilde{x}, \theta, z)$$
$$= |\mu|\exp(i\nu)\frac{-1}{\lambda z}\exp(\mu\tau(z))R\exp(i\theta)$$
$$= R \cdot \exp(\rho\tau(z)) \cdot \frac{-1}{\lambda z} \cdot |\mu| \cdot \exp[i(\omega\tau(z) + \nu + \theta)].$$

Recall the formula for χ from the proof of A4), and the definition of A (before (3.2.11)). Using (3.2.7), we get the following estimate on the partial derivatives of χ:

$$|D_i\mathrm{pr}_j\chi(\tilde{x}, \theta, z)| \leq (1+A)2G_0|D_i(\mathcal{T}_0 \circ \kappa)(\tilde{x}, \theta, z)| \quad (i, j = 1, 2, 3).$$

We use the notation $\mathrm{pr}_{1,2}(\hat{x},\hat{\theta},\hat{z}) := (\hat{x},\hat{\theta})$, and we write '$D_{1,2}$' for the derivative with respect to $\tilde{S}\times\mathbb{R}$ (which corresponds to 'D_1' in Theorem 2.4). From (3.2.22), (3.2.25) and (3.2.11)e) we get

$$|D_3 \mathrm{pr}_{1,2}\mathcal{X}(\tilde{x},\theta,z)|$$

(3.2.26)
$$\leq (1+A)2G_0 \max\{\frac{\tilde{K}(\tilde{R}+\tilde{r})\exp(\gamma\tau(z))}{\lambda|z|}, \frac{R|\mu|\exp(\rho\tau(z))}{\lambda|z|}\}$$
$$= (1+A)\frac{2G_0 R|\mu|\exp(\rho\tau(z))}{\lambda|z|}.$$

(Note that $D_i\mathrm{pr}_3(\mathcal{T}_0\circ\kappa) = 0$, $i=1,2,3$.) From (3.2.21) and (3.2.24) we infer

$$|D_2\mathcal{X}(\tilde{x},\theta,z)| \leq (1+A)2G_0 R\exp(\rho\tau(z)),$$

and from (3.2.20) and (3.2.23) we get

$$|D_1\mathcal{X}(\tilde{x},\theta,z)| \leq (1+A)2G_0 K\exp(\gamma\tau(z)).$$

Thus, using (3.2.11)a), we conclude

(3.2.27)
$$|D_{1,2}\mathcal{X}(\tilde{x},\theta,z)| \leq (1+A)2G_0[R\cdot\exp(\rho\tau(z)) + K\cdot\exp(\gamma\tau(z))]$$
$$\leq (1+A)4G_0 R\cdot\exp(\rho\tau(z)).$$

We now derive a lower bound for $|D_3\mathrm{pr}_3\mathcal{X}(\tilde{x},\theta,z)|$. Assume first that $z>0$. Define $(\tilde{y},w) \in \tilde{S}\times\mathbb{C}$ as above, that is, $(\tilde{y},w) = \mathrm{pr}_{1,2}(\mathcal{T}_0\circ\kappa)(\tilde{x},\theta,z)$. For abbreviation, set

$$l(z) := \frac{-R\exp(\rho\tau(z))|\mu|}{\lambda z}.$$

(Recall that $\mathrm{pr}_3\circ\mathcal{X} = \mathrm{pr}_3\circ\mathcal{T}_1\circ\mathcal{T}_0\circ\kappa$.) Using (3.2.25) and linearity of $DG^+(\tilde{y},w)$ we get

$$D_3\mathrm{pr}_3\mathcal{X}(\tilde{x},\theta,z)(1)$$
$$= \mathrm{pr}_3 DG^+(\tilde{y},w)[D_3(\mathrm{pr}_1\circ\mathcal{T}_0\circ\kappa)(\tilde{x},\theta,z)(1), D_3(\mathrm{pr}_2\circ\mathcal{T}_0\circ\kappa)(\tilde{x},\theta,z)(1)]$$
$$= \mathrm{pr}_3 DG^+(\tilde{y},w)[D_3(\mathrm{pr}_1\circ\mathcal{T}_0\circ\kappa)(\tilde{x},\theta,z)(1), l(z)\exp[i(\omega\tau(z)+\nu+\theta)]]$$
$$= l(z)\mathrm{pr}_3 DG^+(\tilde{y},w)[\frac{1}{l(z)}D_3(\mathrm{pr}_1\circ\mathcal{T}_0\circ\kappa)(\tilde{x},\theta,z)(1), \exp[i(\omega\tau(z)+\nu+\theta)]].$$

One sees from (3.2.22) and from (3.2.11)f) that the first component of the argument of $DG^+(\tilde{y},w)$ satisfies

$$|\frac{1}{l(z)}D_3(\mathrm{pr}_1\circ\mathcal{T}_0\circ\kappa)(\tilde{x},\theta,z)(1)|$$

(3.2.28)
$$\leq \frac{\lambda|z|}{R\cdot\exp(\rho\tau(z))|\mu|}\cdot\frac{\tilde{K}(\tilde{R}+\tilde{r})\exp(\gamma\tau(z))}{\lambda|z|}$$
$$= \frac{\tilde{K}(\tilde{R}+\tilde{r})}{R|\mu|}\exp((\gamma-\rho)\tau(z)) \leq \delta_1.$$

Recall that $\tilde{y}$ and w satisfy

(3.2.29) $$\max\{|\tilde{y}|,|w|\} \leq r_0 \leq \delta_1$$

(compare (3.2.13), (3.2.14)), and that the properties $(\tilde{x},\theta,z) \in \Delta$ and $z > 0$ imply $z \in [a_{l,+},b_{l,+}]$ for an $l \in \{+,-\}$, and $|\theta - \varphi_l| \leq \delta \leq \delta_1/4$. Thus we have (compare (3.2.12)a), b))

(3.2.30) $$\omega\tau(z) + \nu + \theta \in 2\pi k + \theta_{0,+} - \varphi_l + [-3\delta_1/4, 3\delta_1/4] + \nu + \theta$$
$$\subset 2\pi k + \theta_{0,+} + \nu + [-\delta_1,\delta_1].$$

Combining (3.2.28)–(3.2.30) with (3.2.5), we conclude that

$$|\mathrm{pr}_3 DG^+(\tilde{y},w)\bigl[\frac{1}{l(z)}D_3(\mathrm{pr}_1 \circ \mathcal{T}_0 \circ \kappa)(\tilde{x},\theta,z)(1), \exp[i(\omega\tau(z)+\nu+\theta)]\bigr]| \geq c_1.$$

Since κ does not change the third component, it follows that

(3.2.31) $$|D_3\mathrm{pr}_3\chi(\tilde{x},\theta,z)| \geq |l(z)| \cdot c_1 = \frac{R|\mu|c_1 \cdot \exp(\rho\tau(z))}{\lambda|z|}.$$

The proof that the last estimate holds also in case $z < 0$ is analogous. We define
$$L := \frac{(1+A)16 G_0 \lambda B}{|\mu|c_1},$$

and we can now complete the proof of the present lemma by verifying the analogs of the estimates from condition (iii) of Theorem 2.4. It follows from (3.2.12)c) and (3.2.11)g) that

$$L < 1 \text{ and } L < \frac{c_1}{4G_0(1+A)}.$$

In the lines below, the argument $(\tilde{x},\theta,z)$ of χ is omitted. First, using (3.2.31) and (3.2.26), one gets

$$|D_3\mathrm{pr}_3\chi| - L|D_3\mathrm{pr}_{1,2}\chi|$$
$$\geq \frac{R|\mu|c_1}{\lambda} \cdot \frac{\exp(\rho\tau(z))}{|z|} - L \cdot \frac{(1+A)2G_0 R|\mu|}{\lambda} \frac{\exp(\rho\tau(z))}{|z|}$$
$$= \frac{R|\mu|}{\lambda} \frac{\exp(\rho\tau(z))}{|z|}(c_1 - (1+A)2G_0 L)$$
$$\geq \frac{R|\mu|}{\lambda}\frac{c_1}{2}\frac{\exp(\rho\tau(z))}{|z|}$$
$$\geq \min_{z \in \bigcup_{l,m=-,+} J_{l,m}} \frac{\exp(\rho\tau(z))}{|z|} \cdot \frac{R|\mu|}{\lambda}\frac{c_1}{2} > 0,$$

which shows that the first inequality of condition (iii) holds. Further, using (3.2.27) and the second last line from the above estimate, we see that

$$\frac{|D_{1,2}\mathrm{pr}_3\chi| + L|D_{1,2}\mathrm{pr}_{1,2}\chi|}{|D_3\mathrm{pr}_3\chi| - L|D_3\mathrm{pr}_{1,2}\chi|} \leq \frac{(1+L)|D_{1,2}\chi|}{\frac{R|\mu|}{\lambda} \frac{c_1}{2} \frac{\exp(\rho\tau(z))}{|z|}}$$

$$\leq \frac{(1+A)4G_0R\exp(\rho\tau(z))(1+L)}{\frac{R|\mu|}{\lambda} \frac{c_1}{2} \frac{\exp(\rho\tau(z))}{|z|}} = \frac{(1+A)8G_0\lambda}{|\mu|c_1}(1+L)|z|$$

$$< \frac{(1+A)8G_0\lambda}{|\mu|c_1} \cdot 2|z| \leq \frac{(1+A)16G_0\lambda}{|\mu|c_1} \cdot B = L.$$

Thus the second inequality from condition (iii) also holds. Finally, the definition of $B = B^{(k)}$ and of $J_{l,m}^{(k)}$ shows that $B^{(k)} = \exp(-\frac{2\pi\lambda}{\omega}(k-k_0)) \cdot B^{(k_0)}$. Hence, if we put $\tilde{L} := \frac{(1+A)16G_0\lambda}{|\mu|c_1} \cdot B^{(k_0)} \cdot \exp(\frac{2\pi\lambda}{\omega}k_0)$, we have

$$L = L^{(k)} = \frac{16G_0\lambda}{|\mu|c_1} \cdot B^{(k)} = \tilde{L} \cdot \exp(-\frac{2\pi\lambda}{\omega}k),$$

and the last assertion of Lemma 3.2. is proved. □

3.3. Corollary. *Let $k \geq k_0$, $\mathbf{s} = (s_0, s_1, ...) \in \mathcal{S}^+$, and set $m_0 := +$ or $m_0 := -$, according as $s_0 = 1$ or $s_0 = -1$. Let $\varphi_{\mathbf{s}}^{(k)} : I_+ \cup I_- \to J_{+,m_0}^{(k)} \cup J_{-,m_0}^{(k)}$ be as guaranteed by Lemma 3.2 and Theorem 2.4. Then, with $\mathcal{T} := \mathcal{T}_1 \circ \mathcal{T}_0|_{\kappa(\mathcal{D})}$, one has*

$$\kappa(\mathrm{graph}\ (\varphi_{\mathbf{s}}^{(k)})) \subset \mathrm{inv}^+(\mathcal{T}, \kappa(\Delta^{(k)})),$$

and for every $(\tilde{x}, w, z) \in \kappa(\mathrm{graph}\ \varphi_{\mathbf{s}}^{(k)})$ one has $\sigma^+(\tilde{x}, w, z) = \mathbf{s}$. In particular, $\mathcal{T}$ has symbolic dynamics with respect to $\mathcal{S}^+$ on $\kappa(\Delta^{(k)})$ for all $k \geq k_0$.

Proof. We have $\kappa \circ \chi = \mathcal{T} \circ \kappa$ on $\mathcal{D}$. From Theorem 2.4, c), we know that graph $\varphi_{\mathbf{s}}^{(k)} \subset \mathrm{inv}^+(\chi, \Delta^{(k)})$, and $\sigma^+(\tilde{x}, \theta, z) = \mathbf{s}$ for all $(\tilde{x}, \theta, z) \in$ graph $\varphi_{\mathbf{s}}^{(k)}$. Note that κ does not change the third component, and that pr_3 here corresponds to pr_2 in Remark 2.3. The assertion follows from Remark 2.3. □

We can easily extend part of the result of Lemma 3.2 to maps which are nearby in the C^1-sense.

Assume that $k \geq k_0$, that $\bar{\mathcal{D}}_0 \subset \mathcal{D}_0$ is an open neighborhood of $\kappa(\Delta^{(k)})$ in $\tilde{S} \times \mathbb{C} \times \mathbb{R}$, and that $\bar{\Sigma}$ is an open neighborhood of $\{(0,0)\} \times \{-z_0, z_0\}$ in $\Sigma \times \{-z_0, z_0\}$. Consider C^1-maps $\bar{\mathcal{T}}_0 : \bar{\mathcal{D}}_0 \to \bar{\Sigma}$ (here we mean that $\mathrm{pr}_{1,2} \circ \bar{\mathcal{T}}_0$ is C^1) and $\bar{\mathcal{T}}_1 : \bar{\Sigma} \to \tilde{S} \times \mathbb{C}_R \times \mathbb{R}$, such that $\bar{\mathcal{T}} := \bar{\mathcal{T}}_1 \circ \bar{\mathcal{T}}_0$ and $\mathcal{T}$ are defined on $\bar{\mathcal{D}}_0$, and $\bar{\mathcal{T}} - \mathcal{T}|_{\bar{\mathcal{D}}_0}$ is BC^1.

3.4. Corollary. *Let $k \in \mathbb{N}$, $k \geq k_0$. There exists $\varepsilon_k > 0$ such that if a map $\bar{\mathcal{T}} := \bar{\mathcal{T}}_1 \circ \bar{\mathcal{T}}_0$ as above satisfies*

$$\|\bar{\mathcal{T}} - \mathcal{T}_{|\bar{\mathcal{D}}_0}\|_{C^1} < \varepsilon_k$$

then the following statements are true.

1) *There is an open neighborhood U of $\Delta^{(k)}$ in $\tilde{S} \times \mathbb{R} \times \mathbb{R}$ such that $\bar{\mathcal{T}} \circ \kappa$ is defined on U, $\mathrm{image}(\bar{\mathcal{T}} \circ \kappa) \subset \mathrm{image}(\kappa)$, the map*

$$\bar{\chi} := \kappa^{-1} \circ \bar{\mathcal{T}} \circ \kappa_{|U}$$

   is C^1 on U, and the assertions of Theorem 2.4 hold for $\bar{\chi}$ on $\Delta^{(k)}$.

2) *The analog of Corollary 3.3 holds for the functions $\bar{\varphi}_\mathbf{s}^{(k)}$ corresponding to $\bar{\chi}$ and $\mathbf{s} \in \mathcal{S}^+$ with graph $\bar{\varphi}_\mathbf{s}^{(k)} \subset \Delta^{(k)}$. In particular, $\bar{\mathcal{T}}$ also has symbolic dynamics with respect to $\mathcal{S}^+$ on the set $\kappa(\Delta^{(k)})$.*

Proof. Continuity of κ allows us to choose an open neighborhood U of $\Delta^{(k)}$ in $\tilde{S} \times \mathbb{R} \times \mathbb{R}$ with $\kappa(U) \subset \bar{\mathcal{D}}_0$ and $\mathrm{dist}(\mathrm{pr}_3\kappa(U), \{0\}) > 0$. Then $\bar{\mathcal{T}}_1 \circ \bar{\mathcal{T}}_0 \circ \kappa$ is defined on U. In the proof of assertion A3) from Lemma 3.2, we had observed that the following is true for $(\tilde{x}, \theta, z) \in D_+$ and

$$(\hat{x}, \hat{w}, \hat{z}) := (\mathcal{T}_1 \circ \mathcal{T}_0 \circ \kappa)(\tilde{x}, \theta, z) \in \tilde{S} \times \mathbb{C}_R \times \mathbb{R} :$$

With the number $d_1 > 0$ introduced in the passage before formula (3.2.7), one has $\hat{w} + \mathbb{C}(d_1) \subset Q^+$. Further, $\hat{w} = R \cdot \exp(i\hat{\theta})$, with $\hat{\theta} = \arg^+(\hat{w})$, and $(\hat{x}, \hat{\theta}) \in I'_+$. (Corresponding statements hold for $(\tilde{x}, \theta, z) \in D_-$.) Recall that $I'_+ \subset \mathrm{int}_d(I_+)$. It follows that if

$$\|\bar{\mathcal{T}}_1 \circ \bar{\mathcal{T}}_0 - \mathcal{T}_1 \circ \mathcal{T}_{0|\bar{\mathcal{D}}_0}\|_{C^0} < \min\{d_1, d\}$$

then

$$(\bar{x}, \bar{w}, \bar{z}) := (\bar{\mathcal{T}}_1 \circ \bar{\mathcal{T}}_0 \circ \kappa)(\tilde{x}, \theta, z)$$

satisfies $\bar{w} \in Q^+$, $\bar{w} = R \cdot \exp(i\bar{\theta})$ with $\bar{\theta} := \arg^+(\bar{w})$, and $(\bar{x}, \bar{w}) \in I_+$. Thus one obtains $(\bar{x}, \bar{w}, \bar{z}) \in \mathrm{image}(\kappa)$, so

$$(\bar{\mathcal{T}} \circ \kappa)(U) = (\bar{\mathcal{T}}_1 \circ \bar{\mathcal{T}}_0 \circ \kappa)(U) \subset \mathrm{image}(\kappa).$$

It follows now exactly as in the proof of assertion A4) of Lemma 3.2 that $\bar{\chi}$ is C^1 on U.

Recall from step 3 in the proof of Lemma 3.2 that the maps $q^\pm$ and their inverses are Lipschitz with bounded and uniformly continuous derivatives. It follows that κ and κ^{-1} are Lipschitz with bounded and uniformly continuous derivatives. The property $\mathrm{dist}(\mathrm{pr}_3\kappa(U), \{0\}) > 0$ and Remark 3.1, b) together with $G^\pm \in BC^1$ imply that $\mathcal{T} \circ \kappa = \mathcal{T}_1 \circ \mathcal{T}_0 \circ \kappa$ is BC^1 on U.

It follows from uniform continuity of κ^{-1} and of $D\kappa^{-1}$, from $\kappa(U) \subset \bar{\mathcal{D}}_0$ and from Proposition 6.6 that

$$\|(\kappa^{-1} \circ \mathcal{T} - \kappa^{-1} \circ \bar{\mathcal{T}})_{|\kappa(U)}\|_{C^1} \to 0 \text{ as } \|\bar{\mathcal{T}} - \mathcal{T}_{|\bar{\mathcal{D}}_0}\|_{C^1} \longrightarrow 0.$$

Boundedness of $D\kappa$ now shows that

$$\|(\kappa^{-1} \circ \mathcal{T} \circ \kappa - \kappa^{-1} \circ \bar{\mathcal{T}} \circ \kappa)_{|U}\|_{C^1} \to 0 \text{ as } \|\bar{\mathcal{T}} - \mathcal{T}_{|\bar{\mathcal{D}}_0}\|_{C^1} \longrightarrow 0,$$

so $\|\bar{\chi} - \chi_{|U}\|_{C^1} \to 0$ as $\|\bar{\mathcal{T}} - \mathcal{T}_{|\bar{\mathcal{D}}_0}\|_{C^1} \to 0$. Choose $\gamma_1 > 0$ as in Corollary 2.5, applied to $\bar{D} := U$ and $\Delta := \Delta^{(k)}$. There exists $\varepsilon_k \in (0, \min\{d_1, d\})$ such that if $\|\bar{\mathcal{T}} - \mathcal{T}_{|\bar{\mathcal{D}}_0}\|_{C^1} \subset \varepsilon_k$ then $\|\bar{\chi} - \chi_{|U}\|_{C^1} < \gamma_1$. In this case, Corollary 2.5 shows that the statements of Theorem 2.4 hold for $\bar{\chi}$ on $\Delta^{(k)}$. The assertions of 1) are now proved. The proof of the statements under 2) is analogous to the proof of Corollary 3.3. □

4. LINKING EQUATIONS AND MAPS

In this section we establish a connection between the results on maps of Section 3 and the semiflows induced by delay equations.

In order to later apply the robustness statement of Corollary 3.4, we always need to consider not only one particular semiflow, but also nearby ones. Further, we have to deal with nonlinearities which are not everywhere differentiable, in view of our example in Section 5. These are the reasons for a number of technicalities in this section.

We want to describe solutions of delay equations that go 'up and down' between different integer levels. These levels will be encoded by symbol sequences of the form $\mathbf{l} = (...l_{-1}l_0l_1...) \in \mathbb{Z}^{\mathbb{Z}}$.

4.1. Definition. *Let $I \subset \mathbb{R}$ be an interval, and let $x \in C^0(I, \mathbb{R})$. Let $k \in \mathbb{Z}$, and let $J \subset I$ be a subinterval.*

a) We say that x oscillates about k on J if $x(J) \cap \mathbb{Z} = \{k\}$.

b) Let now $k, l \in \mathbb{Z}$ with $k \neq l$. We say that x wanders from k to l on J if there exist subintervals J_k, J_{kl}, J_l of J with $J = J_k \cup J_{kl} \cup J_l$, with $J_k \leq J_{kl} \leq J_l$, and such that x oscillates about k on J_k, x oscillates about l on J_l, and $x(J_{kl}) \cap \mathbb{Z} = \{m \in \mathbb{Z} \mid \min\{k,l\} < m < \max\{k,l\}\}$.

c) Let now a sequence $\mathbf{l} = (...l_{-1}l_0l_1...) \in \mathbb{Z}^{\mathbb{Z}}$ be given. We say that $x \in C^0(\mathbb{R}, \mathbb{R})$ follows the level sequence $\mathbf{l}$ if there exist intervals J_k ($k \in \mathbb{Z}$) with the subsequent properties.

$$\sup_{k \in \mathbb{Z}} (\sup J_k - \inf J_k) < \infty,$$

$$\bigcup_{k \in \mathbb{Z}} J_k = \mathbb{R}, \quad \inf J_k < \inf J_{k+1}, \quad \sup J_k < \sup J_{k+1} \quad (k \in \mathbb{Z}), \text{ and}$$

x wanders from l_k to l_{k+1} on J_k ($k \in \mathbb{Z}$).

The corresponding notion for $x : [-1, \infty) \to \mathbb{R}$ and for level sequences $\mathbf{l} = (l_0, l_1...) \in \mathbb{Z}^{\mathbb{N}_0}$ is defined analogously.

Note that the intervals J_k and J_{k+1} in the above definition may have more than one point in common, and $x(J_k \cap J_{k+1}) \cap \mathbb{Z} \subset \{l_k\}$ ($k \in \mathbb{Z}$). Not also that if x follows $\mathbf{l}$ then one has necessarily $l_k \neq l_{k+1}$ for $k \in \mathbb{Z}$, respectively for $k \in \mathbb{N}_0$.

We can use the above notions to express the existence of a variety of qualitatively different solutions of functional differential equations.

4.2. Definition. *Let $F : C \to \mathbb{R}$ be continuous, and let $\Delta \subset C$ and $\mathcal{L} \subset \mathbb{Z}^{\mathbb{Z}}$ (or $\mathcal{L} \subset \mathbb{Z}^{\mathbb{N}_0}$) be subsets. We say that the functional differential equation*

(F) $$\dot{x}(t) = F(x_t)$$

has symbolic dynamics with respect to $\mathcal{L}$ on Δ if for every $\mathbf{l} \in \mathcal{L}$ there exists a solution $x : \mathbb{R} \to \mathbb{R}$ (or $x : [-1, \infty) \to \mathbb{R}$) of (F) with $x_0 \in \Delta$ which follows the level sequence $\mathbf{l}$.

In Section 5 we construct scalar delay equations of the form

$$(f) \qquad \dot{x}(t) = f(x(t-1))$$

with associated Poincaré-type maps $\mathcal{T}$ to which the results of Section 3 apply. Symbolic dynamics for the map $\mathcal{T}$ then translates into symbolic dynamics in the sense of Definition 4.2 for these equations. The following simple result is a first step towards the construction of such f. It shows that the semigroups generated by linear constant coefficient delay equations satisfy the conditions required in Section 3 (except possibly for condition (T4)).

4.3. Proposition. *Let $\alpha \in (0, 3\pi/2)$, and let $T^\alpha : \mathbb{R}_0^+ \to L_c(C, C)$ be the semigroup defined by the linear constant coefficient equation*

$$(\alpha) \qquad \dot{x}(t) = \alpha x(t-1).$$

There exists a decomposition $C = \tilde{S} \oplus S_2 \times U$ with $\dim S_2 = 2$, $\dim U = 1$, which is invariant under $T^\alpha(t)$ ($t \geq 0$). Every $\psi \in (\tilde{S} \oplus S_2) \setminus \{0\}$ has positive and negative values.

There exists an isomorphism ι from C to the space $X := \tilde{S} \times \mathbb{C} \times \mathbb{R}$ such that the corresponding semigroup $T : \mathbb{R}_0^+ \to L_c(X, X)$, defined by $T(t) = \iota \circ T^\alpha(t) \circ \iota^{-1}$, satisfies conditions (T1)–(T3) of Section 3.

Proof. Consider the complexification $C_\mathbb{C} = \{\varphi + i\psi \mid \varphi, \psi \in C\}$ of C and the corresponding semigroup $T_\mathbb{C}^\alpha : \mathbb{R}_0^+ \to L_c(C_\mathbb{C}, C_\mathbb{C})$,

$$(4.3.1) \qquad T_\mathbb{C}^\alpha(t)(\varphi + i\psi) = T^\alpha(t)\varphi + iT^\alpha(t)\psi \quad (\varphi, \psi \in C).$$

The norm on the complex vector space $C_\mathbb{C}$ is given by

$$|\varphi + i\psi| = \max_{\theta \in [0, 2\pi]} |\cos(\theta) \cdot \varphi + \sin(\theta) \cdot \psi|.$$

The three solutions of the characteristic equation $z = \alpha \exp(-z)$ with largest real part are of the form $\lambda > 0$ and $\mu = \rho + i\omega$, $\bar{\mu} = \rho - i\omega$ with $\rho < 0 < \omega$. (Compare Lemma 6.8, a) in the appendix.) All other solutions z of the characteristic equation satisfy $\mathrm{Re}\, z < \rho$. There exists a decomposition $C_\mathbb{C} = \tilde{S}_\mathbb{C} \oplus \mathbb{C} \cdot \exp(\mu \cdot) \oplus \mathbb{C} \cdot \exp(\bar{\mu} \cdot) \oplus \mathbb{C} \cdot \exp(\lambda \cdot)$ of $C_\mathbb{C}$ into $T_\mathbb{C}^\alpha$-invariant subspaces (see Lemma 6.8, c). (Here the notation $\exp(\lambda \cdot)$ means the corresponding function on $[-1, 0]$.) For $\tilde{v} \in \tilde{S}_\mathbb{C}$ and $\zeta_1, \zeta_2, \zeta_3 \in \mathbb{C}$, $t \geq 0$ one has

$$\begin{aligned}(4.3.2) \qquad & T_\mathbb{C}^\alpha(t)(\tilde{v} + \zeta_1 \exp(\mu \cdot) + \zeta_2 \cdot \exp(\bar{\mu} \cdot) + \zeta_3 \cdot \exp(\lambda \cdot)) \\ & = T_\mathbb{C}^\alpha(t)\tilde{v} + \zeta_1 \cdot \exp(\mu t) \cdot \exp(\mu \cdot) + \zeta_2 \cdot \exp(\bar{\mu} t) \exp(\bar{\mu} \cdot) \\ & \quad + \zeta_3 \cdot \exp(\lambda t) \cdot \exp(\lambda \cdot).\end{aligned}$$

Further, there exist $K > 0$ and $\gamma < 0$ with $\gamma < \rho$ such that

(4.3.3) $$\forall \tilde{v} \in \tilde{S}_{\mathbb{C}} : |T_{\mathbb{C}}^{\alpha}(t)\tilde{v}| \leq K \exp(\gamma t)|\tilde{v}|.$$

The following three spaces define a decomposition of C which is invariant under all $T^{\alpha}(t)$ ($t \geq 0$) (see Lemma 6.8, c):

$$\tilde{S} := \text{Re}(\tilde{S}_{\mathbb{C}}) = \{\varphi \in C \mid \exists \psi \in C : \varphi + i\psi \in \tilde{S}_{\mathbb{C}}\},$$
$$S_2 := \text{Re}(\mathbb{C} \cdot \exp(\mu \cdot) + \mathbb{C} \cdot \exp(\bar{\mu} \cdot)) = \{\zeta \cdot \exp(\mu \cdot) + \bar{\zeta} \cdot \exp(\bar{\mu} \cdot) \mid \zeta \in \mathbb{C}\},$$
$$U := \mathbb{R} \cdot \exp(\lambda \cdot).$$

The projection of $\psi \in C$ onto the last factor in this decomposition is given by $\text{pr}_{\lambda}\psi \exp\lambda(\cdot)$, where

$$\text{pr}_{\lambda}\psi = \frac{1}{1 + \lambda}[\psi(0) + \lambda \int_{-1}^{0} \exp(-\lambda s)\psi(s)ds]$$

(see Lemma 6.8, b)). One sees from this formula that if $\psi \neq 0, \text{pr}_{\lambda}\psi = 0$ then ψ must have values of both signs.

Define

$$\iota : C \to X = \tilde{S} \times \mathbb{C} \times \mathbb{R}, \quad \tilde{v} + \zeta \cdot \exp(\mu \cdot) + \bar{\zeta} \cdot \exp(\bar{\mu} \cdot) + z \cdot \exp(\lambda \cdot) \mapsto (\tilde{v}, \zeta, z).$$

It is obvious that ι is an $\mathbb{R}$–linear homeomorphism. It follows from (4.3.1) and (4.3.2) that the semigroup of the operators $T(t) = \iota \circ T^{\alpha}(t) \circ \iota^{-1}$ ($t \geq 0$) on X is given by

(4.3.4)
$$T(t)(\tilde{v}, \zeta, z)$$
$$= \iota[T_{\mathbb{C}}^{\alpha}(t)\tilde{v} + \zeta \cdot \exp(\mu t) \cdot \exp(\mu \cdot) + \bar{\zeta} \cdot \exp(\bar{\mu} t) \cdot \exp(\bar{\mu} \cdot) + \exp(\lambda t) \cdot \exp(\lambda \cdot)]$$
$$= (T^{\alpha}(t)\tilde{v}, \exp(\mu t) \cdot \zeta, \exp(\lambda t) \cdot z)$$

for $\tilde{v} \in \tilde{S}, \zeta \in \mathbb{C}, z \in \mathbb{R}$. One sees that the decomposition $X = \tilde{S} \times \mathbb{C} \times \mathbb{R}$ is invariant under $T(t)$, so (T1) holds. Further, (T2) holds, in view of (4.3.3) and the above formula, except possibly for the inequality $|\rho| < \lambda$.

Proof of T3: For $t \geq 0$ and $\tilde{v} \in \tilde{S}$, $\tilde{T}(t)\tilde{v}$ is given by the segment $x_t^{\tilde{v}}$, where $x^{\tilde{v}} : [-1, \infty) \to \mathbb{R}$ is the solution of equation (α) with initial segment $\tilde{v}$. The map $(t, \tilde{v}) \mapsto x_t^{\tilde{v}}$ is continuously differentiable on $(1, \infty) \times C$ (compare, e.g., Lemma 1.5 in [Lani-Wayda 3]). In particular, for $\tilde{v} \in \tilde{S}$ the phase curve $[0, \infty) \ni t \mapsto x_t^{\tilde{v}} \in C$ is differentiable at all $t > 1$, and the derivative is given by

$$\frac{d}{ds}[s \mapsto x_s^{\tilde{v}}]\big|_{s=t} = \alpha \cdot x_{t-1}^{\tilde{v}} \quad (t > 1).$$

Setting $t_0 := 2$, we obtain that the map $(t, \tilde{v}) \mapsto \tilde{T}(t)\tilde{v}$ is C^1 on $[t_0, \infty) \times C$. Using (4.3.3), one gets for $t \geq t_0$

$$|\frac{d}{ds}[s \mapsto \tilde{T}(s)\tilde{v}]|_{s=t}| = |\alpha x_{t-1}^{\tilde{v}}| \leq \alpha K \cdot \exp(\gamma(t-1))|\tilde{v}|$$
$$= \alpha \cdot K \cdot \exp(-\gamma)\exp(\gamma t)|\tilde{v}|,$$

so (T3) is proved if we set $\tilde{K} := \alpha K \cdot \exp(-\gamma)$. □

In the present section we describe the connection between equations, semi-flows and maps, and the perturbation results. This part is to some extent independent of the specific form of the feedback function f. However, we have to assume linearity of f in certain regions in order that Proposition 4.3 and the results of Section 3 (which involve semigroups of linear operators) apply. (We comment on these conditions at the end of this section.)

Note that with the functional $\hat{f} : C \to \mathbb{R}$, $\hat{f}(\psi) = f(\psi(-1))$, equation (f) is equivalent to the equation

$$(\hat{f}) \qquad\qquad \dot{x}(t) = \hat{f}(x_t),$$

and thus is embedded in the more general class of equations of type (F). The perturbation results of Section 3 allow us to extend the result on symbolic dynamics to 'nearby' equations of type (F).

Similar to the case of ordinary differential equations, a functional $F \in Lip(C, \mathbb{R})$ generates a (global) semiflow $\mathbb{R}_0^+ \times C \to C$, which we denote by Φ_F. A functional of the form $\hat{f}$ (with $f : \mathbb{R} \to \mathbb{R}$) generates such a semiflow even if f is only continuous. In a slight abuse of notation, we write Φ_f for the semiflow induced by equation (f) (for continuous $f : \mathbb{R} \to \mathbb{R}$), instead of $\Phi_{\hat{f}}$. For $\psi \in C$, we write $y^{\psi, F}$ for the maximal solution of equation (F) with initial segment $y_0^{\psi, F} = \psi$.

We assume the following conditions on f.

(f1) $f : \mathbb{R} \to \mathbb{R}$ is continuous, and the restrictions of f to $[-3/2, -1/2]$ and to $[-1/2, 1/2]$ are C^1. Further, f has the periodicity and oddness properties

$$f(x+2) = f(x) = -f(-x) \quad (x \in \mathbb{R}).$$

(f2) $f(-1) = 0$, and there exist $\alpha \in (0, 3\pi/2)$ and $\bar{l} \in (0, 1/2]$ such that

$$f(-1+x) = \alpha \cdot x \text{ if } |x| \leq \bar{l}.$$

(f3) If $\mu = \rho + i\omega$, $\bar{\mu} = \rho - i\omega$ ($\omega > 0$) and $\lambda > 0$ are the three solutions of the characteristic equation $z = \alpha e^{-z}$ with largest real parts (as in the proof of Proposition 4.3), we assume that

$$|\rho| < \lambda.$$

Note that (f1) and (f2) imply that one has for all $k \in \mathbb{Z}$

(f4) $f(2k - 1 + x) = \alpha x$ if $|x| \le \bar{l}$.

In order to obtain maps $G^\pm$ as in Section 3, which describe a 'global' part of the dynamics of equation (f), it is clear that we need a condition on the global solution behavior. Recall condition (G1) on the maps $G^\pm$, and the maps $\mathcal{T}_1, \mathcal{T}_0$ from Section 3.

Condition (G1) expresses *recurrent* behavior of $\mathcal{T}_1 \circ \mathcal{T}_0$. Below we introduce further assumptions on equation (f), namely existence of *heteroclinic* connections between the levels -1 and $+1$, together with a condition corresponding to property (G2). Due to the periodicity of f, such heteroclinic behavior can be looked upon as a mechanism of recurrence, when all equilibria $2k - 1$ ($k \in \mathbb{Z}$) are identified. Thus the results from Section 3 become applicable.

These ideas are carried out more precisely below. For $\psi \in C$ and $r \in \mathbb{R}$, we shall frequently use the notation $\psi + r$ for the segment defined by $[-1, 0] \ni t \mapsto \psi(t) + r$.

We make the following additional assumptions on f:
There exist a solution $x : \mathbb{R} \to \mathbb{R}$ of (f), numbers $l \in (0, \bar{l})$ and $t_1 > 1$ with the subsequent properties:

(x1) $\forall t \le 0 : |x(t) + 1| \le \bar{l}$.
(x2) $\forall t \in [0, t_1 - 1] : \dot{x}(t) > 0$.
(x3) $\forall t \in [t_1 - 1, \infty) : |x(t) - 1| < l$.
(x4) $\text{pr}_2(x_{t_1} - 1) \ne 0$, where pr_2 denotes the projection on the second factor in the decomposition $C = \tilde{S} \oplus S_2 \oplus U$.

Comments: Note that condition (x1) involves the number $\bar{l}$ from (f2), while condition (x3) uses the new number $l \in (0, \bar{l})$. Note further that assumption (f2) and Proposition 4.3 imply that the unstable space of the linearization of the semiflow Φ_f at the equilibria ± 1 is one–dimensional, namely equal to $U = \mathbb{R} \cdot e^{\lambda \cdot}$. Together with the linearity assumptions on (f), conditions (x1) and (x3) imply that there exists $c \in (0, \bar{l}]$ with

$$x(t) = -1 + c \cdot e^{\lambda t} \text{ for } t \le 0,$$

and that

$$x_t - 1 \in \tilde{S} \oplus S_2 \text{ for all } t \ge t_1, \text{ and } |x_t - 1| \to 0 \ (t \to \infty).$$

(Compare the formula for $T(t)$ in condition (T2).) Thus x is heteroclinic between -1 and 1; see also Figure 6.

Finally, we introduce a condition on the variational equation along x, i.e., the equation

(f, x)
$$\dot{v}(t) = f'(x(t-1))v(t-1).$$

It follows from (f2), from $c \leq \bar{l}$, and from (x4) that $f'(x(t-1)) = \alpha$ for $t \leq 0$ and for $t \geq t_1$. Further, condition (x2) implies that $f'(x(\cdot - 1))$ is defined and continuous at all $t \in \mathbb{R}$ except for two times, where $x(t-1)$ takes the values $-1/2$ and $1/2$. Thus, equation (f,x) is to be understood in the sense of Definition and Remark 6.1 (in the appendix).

Let $\mu = \rho + i\omega$ and $\lambda > 0$ be as in (f3). Conditions (f2) and (x2) imply that $t \mapsto \exp(\mu t)$ is a solution of (f, x) on $(-\infty, 0]$. We assume that the (complex valued) solution $v : [-1, \infty) \to \mathbb{R}$ of equation (f, x) with $v(t) = \exp(\mu t)$ for $t \leq 0$ satisfies

$$(v) \qquad v(t_1) + \lambda \int_{-1}^{0} e^{-\lambda s} v(t_1 + s) ds \neq 0$$

(in other words, $\mathrm{pr}_\lambda v_{t_1} \neq 0$; compare the proof of Proposition 4.3, and Lemma 6.8,b)). Consider now the sets

$$\mathcal{L} := \{(l_k)_{k \in \mathbb{Z}} \in \mathbb{Z}^{\mathbb{Z}} \mid l_0 = -1, |l_{k+1} - l_k| = 2 \ (k \in \mathbb{Z})\}$$

and

$$\mathcal{L}^+ := \{(l_k)_{k \in \mathbb{N}_0} \in \mathbb{Z}^{\mathbb{N}_0} \mid l_0 = -1, |l_{k+1} - l_k| = 2 \ (k \in \mathbb{N}_0)\}.$$

We want to construct solutions of equation (f), and of nearby equations, that follow level sequences from $\mathcal{L}^+$ and from $\mathcal{L}$ in the sense of Definition 4.2. The following proposition is auxiliary in the proof that the finally obtained solutions have the desired oscillation properties.

4.4. Proposition. *There exists $\beta > 0$ such that all segments in the set*

$$(4.4.1) \qquad \mathcal{K} := \{\psi \in C \mid \max\{|\mathrm{pr}_1 \iota \psi|, |\mathrm{pr}_3 \iota \psi|\} \leq \beta |\mathrm{pr}_2 \iota \psi|\}$$

have at least one zero.

Proof. Consider the set

$$M := \{\psi \in C \mid |\mathrm{pr}_1 \iota \psi| = |\mathrm{pr}_3 \iota \psi| = 0, \quad |\mathrm{pr}_2 \iota \psi| = 1\}.$$

Proposition 4.3 shows that every $\psi \in M$ has positive and negative values. Compactness of M implies that there exists $\beta \in (0, 1]$ such that every $\psi \in C$ with

$$(4.4.2) \qquad |\mathrm{pr}_2 \iota \psi| = 1, \quad \max\{|\mathrm{pr}_1 \iota \psi|, |\mathrm{pr}_3 \iota \psi|\} \leq \beta$$

has positive and negative values. If now $\psi \in \mathcal{K} \setminus \{0\}$, then $|\mathrm{pr}_2 \iota \psi| \neq 0$, and $\tilde{\psi} := \dfrac{1}{|\mathrm{pr}_2 \iota \psi|} \psi$ satisfies condition (4.4.2). Consequently, $\tilde{\psi}$ and ψ have a zero. $\square$

Recall the semigroups T^α on C and T on $X = \tilde{S} \times \mathbb{C} \times \mathbb{R}$ with $T^\alpha(t) = \iota^{-1} T(t) \iota$ ($t \geq 0$), and the number $l \in (0, \bar{l})$ from condition (x3), and the fact that $x(t) = -1 + c \cdot \exp(\lambda t)$ for $t \leq 0$. There exist $t_- < -1$ and $r_- > 0$ such that with

$$z_0 := \mathrm{pr}_3 \iota(x_{t_-} + 1) = c \cdot \exp(\lambda t_-) > 0,$$

the following implication holds:

(4.1) $\quad \forall \psi \in C : |\mathrm{pr}_3 \iota \psi| \leq z_0, \quad |\mathrm{pr}_{1,2} \iota \psi| \leq r_- \Rightarrow |\psi| \leq l/2.$

Since, from Proposition 4.3, the semigroup T satisfies conditions (T1)-(T3), there exist numbers $K > 0$ and $\gamma < 0$ as in condition (T2). (Note that $K \geq 1$.) Define

$$U_L := \{ \psi \in C \mid |\mathrm{pr}_3 \iota \psi| < z_0, \; |\mathrm{pr}_{1,2} \iota \psi| < \frac{r_-}{K} \};$$

then $U_L \subset \mathrm{clos}(C(l/2))$. ('L' stands for 'linear'.) For $\psi \in U_L$ with $z := \mathrm{pr}_3 \iota \psi \neq 0$, we set

$$\tau_{z_0}(\psi) := \tau(z) = \frac{1}{\lambda} \log \frac{z_0}{|z|}.$$

4.5. Proposition. *For $\varphi \in -1 + U_L$ with $\mathrm{pr}_3 \iota(\varphi + 1) \neq 0$, and $t \in [0, \tau_{z_0}(\varphi + 1)]$, one has*

$$\Phi_f(t, \varphi) = -1 + (\iota^{-1} T(t) \iota)(\varphi + 1), \text{ and}$$
$$|\Phi_f(t, \varphi) + 1| \leq l/2.$$

Proof. Let $\varphi \in -1 + U_L$ with $\mathrm{pr}_3 \iota(\varphi + 1) \neq 0$, and set $\psi := \varphi + 1$. Then $\tau_{z_0}(\psi) = \tau(\mathrm{pr}_3 \iota \psi)$, and we have the following relations for $t \in [0, \tau_{z_0}(\psi)]$:

$$T(t) \iota \psi = (\tilde{T}(t) \mathrm{pr}_1 \iota \psi, \exp(\mu t) \mathrm{pr}_2 \iota \psi, \exp(\lambda t) z),$$
$$|\mathrm{pr}_3 T(t) \iota \psi| = |\exp(\lambda t) z| \leq z_0,$$
$$|\mathrm{pr}_{1,2} T(t) \iota \psi| \leq \max\{ K \cdot \exp(\gamma t) |\mathrm{pr}_1 \iota \psi|, \exp(\rho t) |\mathrm{pr}_2 \iota \psi| \}$$
$$\leq K \cdot |\mathrm{pr}_{1,2} \iota \psi| \leq r_-.$$

Consequently, $|\iota^{-1} T(t) \iota \psi| \leq l/2$, in view of (4.1). It follows from condition (f2) that $\Phi_f(t, -1 + \psi) = -1 + T^\alpha(t) \psi$. The assertion follows from $T^\alpha(t) = \iota^{-1} T(t) \iota$. □

Define $\mathcal{T}_0 : \tilde{S} \times \mathbb{C} \times [-z_0, z_0] \setminus \{0\} \to X$,

$$\mathcal{T}_0(\tilde{v}, w, z) := T(\tau(z))(\tilde{v}, w, z),$$

as in the passage preceding Remark 3.1. For φ as in Proposition 4.5 we have $|\mathrm{pr}_3 \iota(\varphi + 1)| \in (0, z_0)$ and
(4.2)
$$\Phi_f(\tau_{z_0}(\varphi+1), \varphi) = -1 + (\iota^{-1} \mathcal{T}_0 \iota)(\varphi+1) = -1 + [\iota^{-1} T(\tau(\mathrm{pr}_3 \iota(\varphi+1))) \iota](\varphi+1).$$

We need to extend this relation between Φ_f and $\mathcal{T}_0$ to semiflows induced by equations close to equation (f). Note that the functional $\hat{f} : C \to \mathbb{R}$ induced by f is in $BLip(C, \mathbb{R})$, and that $\hat{f}_{\big| \pm 1 + C(l)} \in BC^1(\pm 1 + C(l), \mathbb{R})$.

4.6. Proposition. *Let $Q \subset C$ be a bounded subset such that*

(4.6.1) $\qquad Q \subset U_L$, and $0 < \inf_{\psi \in Q} |\mathrm{pr}_3 \iota \psi| \leq \sup_{\psi \in Q} |\mathrm{pr}_3 \iota \psi| < z_0 e^{-\lambda}$.

Then there exist an open neighborhood W_Q of Q in C with $W_Q \subset U_L$ and $\beta_Q > 0$ such that with

$$\mathcal{B}_Q := \{F \in BLip(C, \mathbb{R}) \mid F_{\mid -1 + C(l)} \in C^1(-1 + C(l), \mathbb{R}),$$
$$\|(F - \hat{f})_{\mid -1 + C(l)}\|_{C^1} < \beta_Q\}$$

the subsequent statements hold.

a) *For $F \in \mathcal{B}_Q$, there exists a BC^1 map $\tau_{Q,F} : W_Q \to \mathbb{R}$ such that*

$$\tau_{Q, \hat{f}}(\psi) = \tau_{z_0}(\psi) \text{ for } \psi \in W_Q, \text{ and such that}$$

$$\forall \psi \in W_Q \, \forall F \in \mathcal{B}_Q: \quad \mathrm{pr}_3 \iota[\Phi_F(\tau_{Q,F}(\psi), -1 + \psi) + 1] = \mathrm{sign}\,(\mathrm{pr}_3 \iota \psi) \cdot z_0.$$

b) *The maps $P_{Q,F} : W_Q \to C$ defined by*

$$P_{Q,F}(\psi) := \Phi_F(\tau_{Q,F}(\psi), -1 + \psi) + 1$$

are BC^1 and satisfy

$$\|P_{Q,F} - P_{Q,\hat{f}}\|_{C^1} \to 0 \text{ as } \|(F - \hat{f})_{\mid -1 + C(l)}\|_{C^1} \to 0.$$

The derivative $DP_{Q,\hat{f}}$ is uniformly continuous.

c) *For $F \in \mathcal{B}_Q$ and $\psi \in W_Q$, the solution $y^{-1+\psi, F}$ of equation (F) with $y_0^{-1+\psi, F} = -1 + \psi$ satisfies*

$$y^{-1+\psi, F}([-1, \tau_{Q,F}(\psi)]) \cap \mathbb{Z} \subset \{-1\}.$$

Proof. Note that $Q \subset U_L \subset C(l)$. We want to apply Theorem 6.7 from the appendix with $\mathcal{U} := -1 + C(l)$, with $\hat{f}_{\mid -1 + C(l)}$ in place of F, with $\mathcal{D} := -1 + Q$, with $\mathcal{V} := \{\varphi \in \mathcal{U} \mid |\mathrm{pr}_3 \iota(\varphi + 1)| > z_0/2\}$, and with

$$h(\varphi) := \begin{cases} \mathrm{pr}_3 \iota(\varphi + 1) - z_0 & \text{if } \mathrm{pr}_3 \iota(\varphi + 1) > z_0/2, \\ \mathrm{pr}_3 \iota(\varphi + 1) + z_0 & \text{if } \mathrm{pr}_3 \iota(\varphi + 1) < -z_0/2, \end{cases}$$

and with τ_0 defined by $\tau_0(\varphi) := \dfrac{1}{\lambda} \log \dfrac{z_0}{|\mathrm{pr}_3 \iota(\varphi + 1)|} = \tau_{z_0}(\varphi + 1)$ for $\varphi \in \mathcal{D}$. It is clear that $h \in BC^1 Lip(\mathcal{V}, \mathbb{R})$, and the first inequality in (4.6.1) implies that τ_0 is uniformly continuous.

It follows also from (4.6.1) that $1 < \inf_{\varphi \in D} \tau_0(\varphi) \leq \sup_{\varphi \in D} \tau_0(\varphi) < \infty$. Hence there exist $\theta_1 > 0$ and $T > 0$ such that

$$1 < \tau_0(\varphi) - \theta_1 < \tau_0(\varphi) + \theta_1 \leq T \quad (\varphi \in D).$$

We see from Proposition 4.5 and from the obvious estimate

$$|\Phi_f(t,\varphi) - \Phi_f(s,\varphi)| \leq \|f\|_{C^0}|t-s| \quad (\varphi \in C, \ t,s \geq 1)$$

that there exists $\theta_2 > 0$ such that

$$\forall \varphi \in D \ \forall t \in [0, \tau_0(\varphi) + \theta_2] : \ |\Phi_f(t,\varphi) + 1| < 2l/3.$$

We set $\theta_0 := \min\{\theta_1, \theta_2\}$; then conditions (i), (ii) and (v) of Theorem 6.7 hold. It is obvious from the definitions of τ_0 and of $\mathcal{V}$ that condition (vi) is satisfied.

Note that for $\varphi \in D = -1 + Q$, one has $\operatorname{sign} \operatorname{pr}_3 \iota^{-1} T[\tau(\operatorname{pr}_3 \iota(\varphi+1))]\iota(\varphi+1) = \operatorname{sign} \operatorname{pr}_3 \iota(\varphi+1)$. It follows from (4.2) and the definitions of h and τ_0 that

$$\begin{aligned}
h[\Phi_f(\tau_0(\varphi), \varphi)] &= \operatorname{pr}_3 \iota[\Phi_f(\tau_0(\varphi), \varphi) + 1] - \operatorname{sign}(\operatorname{pr}_3 \iota[...]) \cdot z_0 \\
&= \operatorname{pr}_3 \iota[-1 + \iota^{-1} T(\tau_0(\varphi))\iota(\varphi+1) + 1] - \operatorname{sign}(\operatorname{pr}_3 \iota[...]) \cdot z_0 \\
&= \exp(\lambda \tau_0(\varphi)) \operatorname{pr}_3 \iota(\varphi+1) - \operatorname{sign}(\operatorname{pr}_3 \iota(\varphi+1)) \cdot z_0 \\
&= \operatorname{sign}(\operatorname{pr}_3 \iota(\varphi+1)) \cdot z_0 - \operatorname{sign}(\operatorname{pr}_3 \iota(\varphi+1)) \cdot z_0 \\
&= 0, \text{ and}
\end{aligned}$$

$$\begin{aligned}
\Big|\frac{d}{dt}[t \mapsto h(\Phi_{\hat{f}}(t,\varphi))]\big|_{t=\tau_0(\varphi)}\Big| \\
= \Big|\frac{d}{dt}[t \mapsto \exp(\lambda t)\operatorname{pr}_3 \iota(\varphi+1)]\big|_{t=\tau_0(\varphi)}\Big| \\
= |\lambda \exp(\lambda \tau_0(\varphi)) \operatorname{pr}_3 \iota(\varphi+1)| = |\lambda z_0 \operatorname{pr}_3 \iota(\varphi+1)| \\
\geq \lambda z_0 \inf_{\psi \in Q} |\operatorname{pr}_3 \iota \psi| \\
> 0.
\end{aligned}$$

Thus conditions (iii) and (iv) of Theorem 6.7 are also satisfied. Together, we see that Theorem 6.7 can be applied, and we obtain $r > 0$, $\beta > 0$ and the neighborhoods W of $-1 + Q$ in C and $\mathcal{B}$ of $\hat{f}\big|_{-1+C(l)}$ in the space $BC^1(-1+C(l), \mathbb{R})$, and a map $\tau : W \times \mathcal{B} \to \mathbb{R}$ as in Theorem 6.7. The sets W and $\mathcal{B}$ can be chosen such that for $\varphi \in W$, $F \in \mathcal{B}$ one has

(4.6.2) $\quad \operatorname{sign} \operatorname{pr}_3 \iota[\Phi_F(\tau(\varphi, F), \varphi) + 1] = \operatorname{sign} \operatorname{pr}_3 \iota(\varphi+1),$

since this property holds for $F = \hat{f}\big|_{-1+C(l)}$.

Set $W_Q := W + 1$ and $\beta_Q := \beta$. Defining $\mathcal{B}_Q$ as in the statement of the present proposition, it is clear that for $F \in \mathcal{B}_Q$ we have $F\big|_{-1+C(l)} \in \mathcal{B}$, and we can set

$$\tau_{Q,F}(\psi) := \tau(-1+\psi, F\big|_{-1+C(l)}) \text{ for } \psi \in W_Q.$$

Recalling that $\mathcal{U} = -1 + C(l)$, we see from Theorem 6.7, 1) that

$$\tau_{Q,\hat{f}}(\psi) = \tau_0(-1+\psi) = \tau_{z_0}(\psi) \text{ for } \psi \in W_Q, \text{ and}$$

(4.6.3) $\Phi_F(t, -1+\psi) \in -1 + C(l)$ for $F \in \mathcal{B}_Q$, $\psi \in W_Q$, $t \in [0, \tau_{Q,F}(\psi)]$.

Further, for such F and ψ, we have

(4.6.4) $$h(\Phi_F(\tau_{Q,F}(\psi), -1+\psi) = 0.$$

In view of (4.6.2), one gets

(4.6.5)
$$\begin{aligned}&\operatorname{sign} \operatorname{pr}_3 \iota [\Phi_F(\tau_{Q,F}(\psi), -1+\psi) + 1] \\&= \operatorname{sign} \operatorname{pr}_3 \iota [\Phi_F(\tau(-1+\psi, F\big|_{-1+C(l)}), -1+\psi) + 1] \\&= \operatorname{sign} \operatorname{pr}_3 \iota \psi.\end{aligned}$$

The assertions under a) follow from part 1) of Theorem 6.7, and from (4.6.4) and (4.6.5) combined with the definition of h. The assertions under b) follow from part 2) of Theorem 6.7, since $D\hat{f}$ is uniformly continuous (even constant) on $-1 + C(l)$. The property stated in c) is a trivial consequence of (4.6.3) and of $l < 1/2$. $\square$

Note that for Q and W_Q as in Proposition 4.6, and $\psi \in W_Q$, one has from (4.2)
$$(\iota^{-1}\mathcal{T}_0\iota)\psi = \Phi_f(\tau_{z_0}(\psi), -1+\psi) + 1 = P_{Q,\hat{f}}(\psi),$$
and thus
$$\mathcal{T}_0 = \iota P_{Q,\hat{f}} \iota^{-1} \text{ on } \iota(W_Q).$$

For $F \in \mathcal{B}_Q$, we define
$$\mathcal{T}_{0,Q,F} : \iota(W_Q) \to C, \quad \mathcal{T}_{0,Q,F}(\tilde{x}, w, z) := \iota P_{Q,F}(\iota^{-1}(\tilde{x}, w, z)).$$

It follows from Proposition 4.6, b) that

(4.3) $$\|\mathcal{T}_{0,Q,F} - \mathcal{T}_0\big|_{\iota(W_Q)}\|_{C^1} \to 0 \text{ as } \|(F - \hat{f})\big|_{-1+C(l)}\|_{C^1} \to 0.$$

For 2–periodic F (i.e., $F(\psi+2) = F(\psi)$ ($\psi \in C$)), the maps $\mathcal{T}_{0,Q,F}$ describe the 'local' dynamics (close to $2k-1, k \in \mathbb{Z}$) of the functional differential equations that we consider. We now turn to the definition of maps that describe the 'global' dynamics, away from $2k-1$ ($k \in \mathbb{Z}$).

It follows from condition (x4) and the definition of ι that

$$\mathrm{pr}_2 \iota(x_{t_1} - 1) \neq 0.$$

Condition (x3), property (f4) and the formula for $T(t)$ together show that for $t \geq t_1$ one has

(4.4)
$$\begin{aligned}
|\mathrm{pr}_2 \iota(x_t - 1)| &= |\mathrm{pr}_2 \iota T^\alpha(t - t_1)(x_{t_1} - 1)| \\
&= |\mathrm{pr}_2 T(t - t_1) \iota(x_{t_1} - 1)| \\
&= |\exp(\mu(t - t_1))\mathrm{pr}_2 \iota(x_{t_1} - 1)| \\
&= \exp(\rho(t - t_1))|\mathrm{pr}_2 \iota(x_{t_1} - 1)| \\
&> 0.
\end{aligned}$$

Further, for $t \geq t_1$, one has $|\mathrm{pr}_1 \iota(x_t - 1)| \leq K \exp(\gamma(t - t_1))|\mathrm{pr}_1 \iota(x_{t_1} - 1)|$. Recall $\beta \in (0, 1]$ from Proposition 4.4, and that $x_t \to 1$ ($t \to \infty$). There exists $t_+ > t_1 + 1$ with the subsequent properties:

(t_+,i) $\qquad |\mathrm{pr}_1 \iota(x_{t_+} - 1)| < \dfrac{\beta}{2} |\mathrm{pr}_2 \iota(x_{t_+} - 1)|,$

(t_+,ii) $\qquad x_{t_+} - 1 \in U_L,$

(t_+,iii) $\qquad \beta |\mathrm{pr}_2 \iota(x_{t_+} - 1)| < z_0, \quad |\mathrm{pr}_2 \iota(x_{t_+} - 1)| < r_-/K.$

We set $R := |\mathrm{pr}_2 \iota(x_{t_+} - 1)| > 0$, and we first define maps that describe solutions close to $x_{|[t_1, t_+]}$.

4.7. Proposition. *There exist a neighborhood $U_1^+ \subset 1 + C(l)$ of x_{t_1} in C, an open interval $I_R^+ \subset \mathbb{R}$ with $t_+ - t_1 \in I_R^+$, and $\beta_1^+ > 0$ such that with*

$$\mathcal{B}_1^+ := \{ F \in BLip(C, \mathbb{R}) \mid F_{|1 + C(l)} \in BC^1, \|(F - \hat{f})_{|1 + C(l)}\|_{C^1} < \beta_1^+ \},$$

the following statements hold.

a) *For $F \in \mathcal{B}_1^+$, there exists a BC^1 map $\tau_{R,F}^+ : U_1^+ \to I_R^+ \subset \mathbb{R}$ such that*

$$\forall \psi \in U_1^+ \, \forall F \in \mathcal{B}_1^+ \, \forall t \in I_R^+ :$$
$$|\mathrm{pr}_2 \iota(\Phi_F(t, \psi) - 1)| = R \Leftrightarrow t = \tau_{R,F}^+(\psi).$$

b) *The maps $P_{R,F}^+ : U_1^+ \to C$ defined by*

$$P_{R,F}^+(\psi) := \Phi_F(\tau_{R,F}^+(\psi), \psi) \quad (\text{for } F \in \mathcal{B}_1^+)$$

are BC^1 and satisfy

$$\|P^+_{R,F} - P^+_{R,\hat{f}}\|_{C^1} \to 0 \text{ as } \|(F - \hat{f})_{|1+C(l)}\|_{C^1} \to 0,$$

and $DP^+_{R,\hat{f}}$ is uniformly continuous.

c) If $\psi \in U^+_1$, $F \in \mathcal{B}^+_1$, then $y^{\psi,F}([-1, \tau^+_{R,F}(\psi)]) \cap \mathbb{Z} = \{1\}$.

Proof. Set $\psi_1 := x_{t_1}$, and set $\tilde{x} := x(t_1 + \cdot) : \mathbb{R} \to \mathbb{R}$, so $\tilde{x}_0 = \psi_1$. From condition (x3), we have $\tilde{x}_t \in 1+C(l)$ for all $t \geq 0$. Since $x_{t_1} - 1 \in \tilde{S} \oplus S_2$, and $x_{t_1} - 1 \neq 0$ (compare condition (x4)), Proposition 4.3 implies the following fact:

(4.7.1) $\qquad\qquad \psi_1 - 1$ has positive and negative values.

Note now that we have

$$|\text{pr}_2\iota(\tilde{x}_{t_+ - t_1} - 1)| = |\text{pr}_2\iota(x_{t_+} - 1)| = R,$$

and, from (4.4),
(4.7.2)
$$\frac{d}{dt}[t \mapsto |\text{pr}_2\iota(\tilde{x}_t - 1)|]\big|_{t = t_+ - t_1} = \rho \exp(\rho(t_+ - t_1))|\text{pr}_2\iota(x_{t_1} - 1)| < 0.$$

Define
$$h : C \to \mathbb{R}, \quad h(\psi) := |\text{pr}_2\iota(\psi - 1)| - R.$$

Then $h(\psi) = \sqrt{\text{pr}_2\iota(\psi - 1)\overline{\text{pr}_2\iota(\psi - 1)}} - R$, and $Dh(\psi)$ exists for all $\psi \in C$ with $\text{pr}_2\iota(\psi - 1) \neq 0$, and for $\varphi \in C$ one has

$$Dh(\psi)\varphi = \frac{1}{2(h(\psi) + R)}[\text{pr}_2\iota\varphi \cdot \overline{\text{pr}_2\iota(\psi - 1)} + \text{pr}_2\iota(\psi - 1)\overline{\text{pr}_2\iota\varphi}]$$
$$= \frac{1}{h(\psi) + R}\text{Re}[\overline{\text{pr}_2\iota\varphi} \cdot \text{pr}_2\iota(\psi - 1)].$$

This formula shows that Dh is (even uniformly) continuous on the set $\mathcal{V} := \{\psi \in C \mid h(\psi) + R > R/2\}$, which contains $\tilde{x}_{t_+ - t_1}$. We want to apply Theorem 6.7 in the 'one-point-case', with the singleton $D := \{x_{t_1}\}$, with $\mathcal{U} := 1 + C(l)$, and with $\mathcal{V}$ and h as above. Setting $\tau_0(x_{t_1}) := t_+ - t_1 > 1$, it is trivial that we can find $\theta_0, T > 0$ such that all conditions of Theorem 6.7, except for condition (iv), are satisfied. The latter condition holds in view of (4.7.2), with $d_0 := |\rho| \exp(\rho(t_+ - t_1))|\text{pr}_2\iota(x_{t_1} - 1)|$. Applying Theorem 6.7 we obtain a neighborhood $\tilde{U}_1$ of ψ_1 in $\mathcal{U}$, an open interval $I^+_R \subset \mathbb{R}$ with $t_+ - t_1 \in I^+_R$, a number $\beta^+_1 > 0$ such that with

$$\tilde{\mathcal{B}}_1 := \{\tilde{F} \in BC^1(1 + C(l), \mathbb{R}) \mid \|\tilde{F} - \hat{f}_{|1+C(l)}\|_{C^1} < \beta^+_1\},$$

the following hold:

One has a BC^1 map $\tau_R : \tilde{U}_1 \times \tilde{\mathcal{B}}_1 \to I_R^+$ such that for $\psi \in \tilde{U}_1$ and $\tilde{F} \in \tilde{\mathcal{B}}_1$, the solution $y^{\psi,\tilde{F}}$ exists on $[-1, \tau_R(\psi, \tilde{F})]$ and satisfies

$$(4.7.3) \qquad \forall\, t \in [0, \tau_R(\psi, \tilde{F})] : \quad y_t^{\psi,\tilde{F}} \in 1 + C(l).$$

In view of (4.7.1) and $\tilde{U}_1 \subset 1+C(l)$, we can choose a neighborhood $U_1^+ \subset \tilde{U}_1$ of ψ_1 with the property

$$(4.7.4) \qquad \forall\, \psi \in U_1^+ : \quad \psi([-1,0]) \cap \mathbb{Z} = \{1\}.$$

With β_1^+ from above, define now $\mathcal{B}_1^+$ as in the statement of the proposition. Then for $F \in \mathcal{B}_1^+$ we have $\tilde{F} := F\big|_{1+C(l)} \in \tilde{\mathcal{B}}_1$, and we set

$$\tau_{R,F}^+(\psi) := \tau_R(\psi, \tilde{F}) \text{ for } \psi \in U_1^+ \text{ and } F \in \mathcal{B}_1^+.$$

It follows then from (4.7.3) and from Theorem 6.7 that for all $\psi \in U_1^+$ and $F \in \mathcal{B}_1^+$ the statements of a) and b) are true. (Note that $D\hat{f}$ is uniformly continuous on $1+C(l)$, even constant.) Finally, one sees from (4.7.3) and (4.7.4) that statement c) also holds. □

We now construct symmetric counterparts to U_1^+ and to the maps from Proposition 4.7. Recall that $x(t) = -1 + ce^{\lambda t}$ for $t \leq 0$, with some $c \in (0, \bar{l}]$. Oddness of f implies that $\tilde{x} := -x$ is a heteroclinic solution of equation (f) joining 1 to -1. There exists a neighborhood $U_1^- \subset 1+C(l)$ of $\tilde{x}_{t_1}$ in C and a number $\beta_1^- > 0$ such that with

$$\mathcal{B}_1^- := \{F \in BLip(C, \mathbb{R}) \mid F\big|_{-1+C(l)} \in BC^1,$$
$$\|(F - \hat{f})\big|_{-1+C(l)}\|_{C^1} < \beta_1^-\},$$

there exist an open interval $I_R^- \subset \mathbb{R}$ and, for every $F \in \mathcal{B}_1^-$, maps

$$\tau_{R,F}^- : U_1^- \to I_R^-, \quad P_{R,F}^- : U_1^- \to C$$

with properties analogous to those listed in Proposition 4.7. There exists an open, convex neighborhood U_1 of x_{t_1} in C such that, with U_1^+ from Proposition 4.7, one has

$$U_1 \subset U_1^+ \text{ and } -U_1 \subset U_1^-.$$

For $F \in \mathcal{B}_1^+ \cap \mathcal{B}_1^-$, we define

$$P_{R,F} : U_1 \cup (-U_1) \to C, \quad P_{R,F} := P_{R,F}^\pm \text{ on } \pm U_1.$$

From Proposition 4.7, from the analogous statements for the maps $P^-_{R,F}$, from convexity of U_1, and from the fact that $\mathrm{dist}(U_1, -U_1) \geq 2(1-l) > 0$ we infer that for $F \in \mathcal{B}^+_1 \cap \mathcal{B}^-_1$ the map $P_{R,F}$ is BC^1, uniformly continuous and has a uniformly continuous derivative.

Now we turn to the definition of maps that describe solutions in the neighborhood of $x_{|[t_-,t_1]}$, i.e., along the 'global' part of the heteroclinic solution x.

The fact that $x(t) = -1 + ce^{\lambda t}$ for $t \leq 0$ and property (x2) together imply that
$$d_1 := \frac{1}{2}\min\{\dot{x}(t) \mid t \in [t_-, t_1 - 1]\} > 0.$$

Further, properties (x1)–(x3) show that $x(0) \leq -1/2 < 0 < 1/2 < x(t_1 - 1)$. Recall (4.7.1), and that $t_- < -1$. It follows from Lemma 6.3 in the appendix (in particular, from part b)) that we can choose a convex, open, bounded neighborhood U_- of x_{t_-} in C with $U_- \cap (-U_-) = \emptyset$, and a neighborhood $\mathcal{B}_-$ of $\hat{f}$ in $(BLip(C, R), \|\ \|_{C^0})$ such that for all $\psi \in U_- \cup (-U_-)$ and all $F \in \mathcal{B}_-$, the subsequent properties (4.5) - (4.9) hold.

(4.5) $\Phi_F(t_1 - t_-, \psi) \in \pm U_1$ according as $\psi \in \pm U_-$.

(4.6) $\begin{cases} y^{\psi, F}([-1, t_1 - t_- - 1]) \cap \mathbb{Z} = \{0\}, \text{ and} \\ y^{\psi, F}([t_1 - t_- - 1, t_1 - t_-]) \cap \mathbb{Z} = \{\pm 1\} \text{ according as } \psi \in \pm U_-. \end{cases}$

(4.7) $y^{\psi, F}([-1, t_1 - t_-]) \cap (\{-1/2\} + \mathbb{Z}) = \{-1/2, 1/2\}$.

(4.8) $\forall F \in \mathcal{B}_- \ \forall t \in [0, t_1 - t_- - 1]$:
$$\forall \psi \in U_-: \ \dot{y}^{\psi, F}(t) \geq d_1, \quad \forall \psi \in (-U_-): \ \dot{y}^{\psi, F}(t) \leq -d_1.$$

(4.9) For $\psi \in U_- \cup (-U_-)$, there exist unique times
$\tau_{-1/2}(\psi), \tau_{1/2}(\psi) \in (0, t_1 - t_-)$ such that $y^{\psi, \hat{f}}(\tau_{\pm 1/2}(\psi) - 1) = \pm 1/2$, and

$$1 < \inf_{\psi \in U_-} \tau_{-1/2}(\psi) \leq \sup_{\psi \in U_-} \tau_{-1/2}(\psi) < \inf_{\psi \in U_-} \tau_{1/2}(\psi) \leq \sup_{\psi \in U_-} \tau_{1/2}(\psi)$$
$$< t_1 - t_-,$$
$$1 < \inf_{\psi \in (-U_-)} \tau_{1/2}(\psi) \leq \sup_{\psi \in (-U_-)} \tau_{1/2}(\psi) < \inf_{\psi \in (-U_-)} \tau_{-1/2}(\psi)$$
$$\leq \sup_{\psi \in (-U_-)} \tau_{-1/2}(\psi) < t_1 - t_-.$$

(Recall that the points $\pm 1/2$ are the points where f is possibly not differentiable.)

It follows from Lemma 6.5, b) (in the appendix) that we can, in addition, choose U_- such that the map $P_{1,\hat{f}} := \Phi_{\hat{f}}(t_1 - t_-, \cdot)$ is C^1 on $U_- \cup (-U_-)$, with $DP_{1,\hat{f}}$ uniformly continuous. Further, for $\psi \in U_- \cup (-U_-)$, Lemma 6.5,

c) shows that $DP_{1,\hat{f}}(\psi)$ is given by solutions of the variational equation of f along $y^{\psi,\hat{f}}$, in the sense of the definition in 6.1. Boundedness and convexity of U_- imply that
$$P_{1,\hat{f}} \in BC^1(U_- \cup (-U_-), C),$$
and that $P_{1,\hat{f}}$ is uniformly continuous.

For $F \in \mathcal{B}_-$, set
$$P_{1,F} : U_- \cup (-U_-) \to C, \quad P_{1,F}(\psi) := \Phi_F(t_1 - t_-, \psi).$$

One sees from (4.5) that these maps are bounded. It follows from $t_1 - t_- > 1$ and from Lemma 1.5, b) in [Lani-Wayda 3] that for $F \in \mathcal{B}_- \cap BC^1(C, \mathbb{R})$, the map $P_{1,F}$ is C^1. Now boundedness of U_-, and the estimate from Proposition 6.2 applied to the solutions of variational equations along solutions of equation (F), show that
$$\forall F \in \mathcal{B}_- \cap BC^1(C, \mathbb{R}): \quad P_{1,F} \in BC^1(U_- \cup (-U_-), C).$$

Next, we need to investigate closeness of the maps $P_{1,F}$ to $P_{1,\hat{f}}$. The perturbation result in Proposition 4.8 below is not quite standard, and not included in the results in the C^1–setting from [Lani-Wayda 3], because the heteroclinic solution x crosses the values where f may be nondifferentiable. This technical fact is also the reason why we did not combine the maps $P_{1,F}$ and $P_{R,F}^{\pm}$ to one single map (for each F) – thus the perturbation statement in Proposition 4.8 below is needed only for the time $t_1 - t_-$ maps $P_{1,F}$ (with *fixed* time $t_1 - t_-$).

Note that for $\psi \in C$ with $\psi(-1) \notin -1/2 + \mathbb{Z}$, the derivative $D\hat{f}(\psi)$ exists, and $D\hat{f}(\psi)\chi = f'(\psi(-1))\chi(-1)$ for $\chi \in C$. It follows from (4.9) that for $\psi \in U_- \cup (-U_-)$ the function $[0, t_1 - t_-] \setminus \{\tau_{-1/2}(\psi), \tau_{1/2}(\psi)\} \ni t \mapsto D\hat{f}(y_t^{\psi,\hat{f}})$ in the integrand above is piecewise continuous in the sense defined at the beginning of the appendix. The integral in the proposition below is thus well-defined.

4.8. Proposition. *For every $\varepsilon > 0$ there exists $\theta > 0$ such that for all $F \in BC^1(C, \mathbb{R}) \cap \mathcal{B}_-$ which satisfy the conditions*

(i, θ) $\|F - \hat{f}\|_{C^0} \leq \theta$,

(ii, θ) $\forall \psi \in U_- \cup (-U_-) : \int_0^{t_1 - t_-} |DF(y_t^{\psi,F}) - D\hat{f}(y_t^{\psi,\hat{f}})| dt \leq \theta$,

and

(4.8.1) $$\|DF\|_\infty \leq \|D\hat{f}\|_\infty + 1,$$

one has $\|P_{1,F} - P_{1,\hat{f}}\|_{C^1} \leq \varepsilon$. (Here $\|\ \|_{C^1}$ means the C^1–norm on the space $BC^1(U_- \cup (-U_-), C)$.)

Proof. Let $\varepsilon > 0$. Clearly, $BC^1(C, \mathbb{R}) \subset BLip(C, \mathbb{R})$. It follows from estimate (6.3.1) of Lemma 6.3 that there exists $\theta_1 > 0$ such that for $F \in BC^1(C, \mathbb{R}) \cap \mathcal{B}_-$ with $\|F - \hat{f}\|_{C^0} \leq \theta_1$, one has

$$\|P_{1,F} - P_{1,\hat{f}}\|_{C^0} \leq \varepsilon. \tag{4.8.2}$$

Let now $\psi \in U_- \cup (-U_-)$ and $\varphi \in C$. For $F \in BC^1(C, \mathbb{R})$, the segment $DP_{1,F}(\psi)\varphi$ is given by $v^{\psi,\varphi}_{t_1 - t_-}$, where $v^{\psi,\varphi} : [-1, \infty) \longrightarrow \mathbb{R}$ is the solution of $v_0 = \varphi$, $\dot{v}(t) = DF(y^{\psi,F}_t)v_t$ ($t \geq 0$). (See [Hale, Verduyn Lunel], p.49, Thm. 4.1.) We know from the construction of U_- and from Lemma 6.5 that $DP_{1,\hat{f}}(\psi)$ exists and $DP_{1,\hat{f}}(\psi)\varphi$ is given by $w^{\psi,\varphi}_{t_1 - t_-}$, where $w^{\psi,\varphi}$ is the solution of $w_0 = \varphi$, $\dot{w}(t) = D\hat{f}(y^{\psi,\hat{f}}_t)w_t$ in the sense defined in the appendix (see 6.1 and Lemma 6.5, c).) Thus we have

$$|DP_{1,F}(\psi)\varphi - DP_{1,\hat{f}}(\psi)\varphi| = |v^{\psi,\varphi}_{t_1 - t_-} - w^{\psi,\varphi}_{t_1 - t_-}|. \tag{4.8.3}$$

Set $\nu_1(\psi, F) := \int_0^{t_1 - t_-} |DF(y^{\psi,F}_t) - D\hat{f}(y^{\psi,\hat{f}}_t)| dt$ (for $F \in BC^1(C, \mathbb{R}) \cap \mathcal{B}_-$ and $\psi \in U_- \cup (-U_-)$), and set $\nu_2(\psi, F) := \int_0^{t_1 - t_-} |DF(y^{\psi,F}_t)| dt$. If F satisfies (4.8.1) then, for all $\psi \in U_- \cup (-U_-)$,

$$|\nu_2(\psi, F)| \leq (t_1 - t_-)(\|D\hat{f}\|_\infty + 1).$$

From Proposition 6.2 we obtain the estimate

$$|v^{\psi,\varphi}_{t_1 - t_-} - w^{\psi,\varphi}_{t_1 - t_-}| \leq |\varphi|\nu_1(\psi, F)\exp(\nu_2(\psi, F))\exp[\|D\hat{f}(y^{\psi,\hat{f}}_{(\cdot)})\|_\infty(t_1 - t_-)]$$
$$\leq |\varphi|\nu_1(\psi, F)\exp[(t_1 - t_-)(2\|D\hat{f}\|_\infty + 1)].$$

Set $\theta_2 := \varepsilon / \exp[(t_1 - t_-)(2\|D\hat{f}\|_\infty + 1)]$ and $\theta := \min\{\theta_1, \theta_2\}$. For F as in the statement of the proposition, the above estimate, the inequality $\nu_1(\psi, F) \leq \theta_2$, and (4.8.3) show that

$$\|DP_{1,F} - DP_{1,\hat{f}}\|_{C^0} \leq \varepsilon.$$

Together with (4.8.2), the assertion follows. $\square$

Condition (ii, θ) in the above proposition is still not convenient to verify. We show that it holds for suitable perturbations of equation (f) of the form $\dot{x}(t) = g(x(t-1))$.

4.9. Remark. *For every $\theta > 0$ there exist $\theta_1, \theta_2 > 0$ with the following property.*

If $g \in BC^1(\mathbb{R}, \mathbb{R})$ satisfies the inequalities

$$\|g'\|_{C^0} \leq \sup_{x \in \mathbb{R} \setminus (1/2 + \mathbb{Z})} |f'(x)| + 1,$$

$$\|g - f\|_{C^0} \leq \theta_1,$$

$$\|(g - f)|_{\mathbb{R} \setminus (1/2 + (-\theta_2, \theta_2) + \mathbb{Z})}\|_{C^1} \leq \theta_1,$$

then the functional $\hat{g} : C \to \mathbb{R}$, $\hat{g}(\psi) = g(\psi(-1))$ satisfies the conditions of Proposition 4.8 with the given number θ.

Proof. (For $g \in BLip(\mathbb{R}, \mathbb{R})$ and $\psi \in C$, we write $y^{\psi, g}$ instead of $y^{\psi, \hat{g}}$.)

Note that $\|D\hat{f}\|_\infty = \sup_{x \in \mathbb{R} \setminus (1/2 + \mathbb{Z})} |f'(x)| = \|f'\|_\infty$. Recall the number $d_1 > 0$ and properties (4.8), (4.9) from the choice of U_-.

Let $\theta > 0$. Choose $\theta_3 > 0$ with the subsequent properties.

1) $d_1 \theta_3 \leq 1/4$.
2) $1 + l < 3/2 - \theta_3 d_1/2$.
3) For all $\psi \in U_-$, one has

$$1 < \tau_{-1/2}(\psi) - \theta_3 < \tau_{-1/2}(\psi) + \theta_3 < \tau_{1/2}(\psi) - \theta_3 < \tau_{1/2}(\psi) + \theta_3 < t_1 - t_-,$$

and the analogous estimates for $\psi \in (-U_-)$, with $\tau_{-1/2}(\psi)$ and $\tau_{1/2}(\psi)$ interchanged.

4) $4\theta_3 (2\|f'\|_\infty + 1) \leq \theta/3$.

Then properties (4.6) and (4.8), together with (4.5) and the inclusion $U_1 \subset 1 + C(l)$, imply the following inclusions for $\psi \in U_-$:

$$y^{\psi, f}(t - 1) \in (-1, -1/2 - d_1 \theta_3] \text{ if } t \in [0, \tau_{-1/2}(\psi) - \theta_3],$$

$$y^{\psi, f}(t - 1) \in [-1/2 + d_1 \theta_3, 1/2 - d_1 \theta_3] \text{ if } t \in [\tau_{-1/2}(\psi) + \theta_3, \tau_{1/2}(\psi) - \theta_3],$$

$$y^{\psi, f}(t - 1) \in [1/2 + d_1 \theta_3, 1 + l] \text{ if } t \in [\tau_{1/2}(\psi) + \theta_3, t_1 - t_-].$$

The analogous inclusions hold for $\psi \in (-U_-)$, with the intervals containing $y^{\psi, F}(t - 1)$ replaced by their negatives, and with $\tau_{-1/2}(\psi)$ and $\tau_{1/2}(\psi)$ interchanged. Denote the three t–intervals above by $I_1(\psi), I_2(\psi)$, and $I_3(\psi)$. It follows from estimate (6.3.1) of Lemma 6.3 that there exists $\theta_4 > 0$ such that for all $g \in BLip(\mathbb{R}, \mathbb{R})$ with $\|g - f\|_{C^0} \leq \theta_4$, one has for $\psi \in \pm U_-$ the inclusions

(4.9.1)
$$y^{\psi, g}(t - 1) \in \pm[-5/4, -1/2 - d_1 \theta_3/2] \text{ for } t \in I_1(\psi),$$
$$y^{\psi, g}(t - 1) \in \pm[-1/2 + d_1 \theta_3/2, 1/2 - d_1 \theta_3/2] \text{ for } t \in I_2(\psi),$$
$$y^{\psi, g}(t - 1) \in \pm[1/2 + d_1 \theta_3/2, 3/2 - d_1 \theta_3/2] \text{ for } t \in I_3(\psi),$$

where '+' stands in case $\psi \in U_-$ and '−' in case $\psi \in (-U_-)$. (For the third inclusion, recall property 2) of θ_3.) We abbreviate the three intervals in (4.9.1) (for the case '+') by J_1, J_2 and J_3, and we set

$$\theta_2 := d_1 \theta_3 / 2.$$

The set

$$J := J_1 \cup J_2 \cup J_3 \cup (-J_1) \cup (-J_2) \cup (-J_3)$$

satisfies $J \subset \mathbb{R} \setminus (1/2 + (-\theta_2, \theta_2) + \mathbb{Z})$. It follows from property (f1) that f is C^1 on J, so f' is uniformly continuous on J, and there exists $\theta_5 > 0$ such that one has

$$(4.9.2) \qquad \forall \xi, \eta \in J, \ |\xi - \eta| \leq \theta_5 : \quad |f'(\xi) - f'(\eta)| \leq \frac{\theta}{3(t_1 - t_-)}.$$

Using estimate (6.3.1) again, one sees that there exists $\theta_6 > 0$ such that for $g \in BLip(C, \mathbb{R})$ with $\|g - f\|_{C^0} \leq \theta_6$ and for $\psi \in U_- \cup (-U_-)$ one has

$$(4.9.3) \qquad \forall t \in [0, t_1 - t_-]: \quad |y^{\psi,g}(t-1) - y^{\psi,f}(t-1)| \leq \theta_5.$$

Set $\theta_1 := \min\{\theta_6, \theta, \theta_4, \theta/3(t_1 - t_-)\}$. Consider now $g \in BC^1(\mathbb{R}, \mathbb{R})$ with

$$(4.9.4) \qquad \|g'\|_{C^0} \leq \|f'\|_\infty + 1, \quad \|g - f\|_{C^0} \leq \theta_1,$$

and with $\|(g-f)_{|\mathbb{R} \setminus (1/2 + (-\theta_2, \theta_2) + \mathbb{Z})}\|_{C^1} \leq \theta_1$. It follows from (4.9.4) and from $\theta_1 \leq \theta$ that $\hat{g}$ satisfies condition (i, θ) of Proposition 4.8 and condition (4.8.1). Further, for $\psi \in U_- \cup (-U_-)$ and $t \in I_1(\psi) \cup I_2(\psi) \cup I_3(\psi)$, it follows from $\theta_1 \leq \theta_4$ and from (4.9.1) that $y^{\psi,g}(t-1) \in J$ and $y^{\psi,f}(t-1) \in J$. We see from $\theta_1 \leq \theta_6$ that estimate (4.9.3) holds. Thus, in view of the properties (4.9.2) and $J \subset \mathbb{R} \setminus (1/2 + (-\theta_2, \theta_2) + \mathbb{Z})$, we have for $t \in I_1(\psi) \cup I_2(\psi) \cup I_3(\psi)$:

$$|g'(y^{\psi,g}(t-1)) - f'(y^{\psi,f}(t-1))|$$
$$\leq |g'(y^{\psi,g}(t-1)) - f'(y^{\psi,g}(t-1))| + |f'(y^{\psi,g}(t-1)) - f'(y^{\psi,f}(t-1))|$$
$$\leq \|(g' - f')_{|\mathbb{R} \setminus (1/2 + (-\theta_2, \theta_2) + \mathbb{Z})}\|_{C^0} + \frac{\theta}{3(t_1 - t_-)}$$
$$\leq \theta_1 + \frac{\theta}{3(t_1 - t_-)}.$$

Note that $[0, t_1 - t_-] \setminus (I_1(\psi) \cup I_2(\psi) \cup I_3(\psi))$ is the union of two intervals

of length $2\theta_3$. Consequently,

$$\int_0^{t_1-t_-} |D\hat{g}(y_t^{\psi,g}) - D\hat{f}(y_t^{\psi,f})| dt$$

$$= \int_{[0,t_1-t_-]\setminus\{\tau_{-1/2}(\psi),\tau_{1/2}(\psi)\}} |g'(y^{\psi,g}(t-1)) - f'(y^{\psi,f}(t-1))| dt$$

$$\leq \int_{I_1(\psi) \cup I_2(\psi) \cup I_3(\psi)} \ldots + 2 \cdot 2\theta_3 (\|g'\|_{C^0} + \|f'\|_\infty)$$

$$\leq (t_1 - t_-)(\theta_1 + \frac{\theta}{3(t_1 - t_-)}) + 4\theta_3(\|g'\|_{C^0} + \|f'\|_\infty)$$

$$\leq \theta/3 + \theta/3 + 4\theta_3(2\|f'\|_\infty + 1)$$

$$\leq \theta.$$

Hence $\hat{g}$ also satisfies condition (ii, θ) of Proposition 4.8. □

We choose $r_\Sigma > 0$ such that the set

$$\Sigma_C := x_{t_-} + \{\psi \in C \mid \text{pr}_3 \iota \psi = 0, |\text{pr}_{1,2} \iota \psi| < r_\Sigma\}$$

satisfies $\Sigma_C \subset U_-$. Set

$$\Sigma := \{(\tilde{v}, w) \in \tilde{S} \times \mathbb{C} \mid \max\{|\tilde{v}|, |w|\} < r_\Sigma\}.$$

Note that $\iota(x_{t_-} + 1) = (0, 0, z_0)$, and hence one has, for $(\tilde{v}, w) \in \Sigma$,

$$x_{t_-} + \iota^{-1}(\tilde{v}, w, 0) = -1 + \iota^{-1}(\tilde{v}, w, z_0) \in \Sigma_C \subset U_-.$$

For all F satisfying

(4.10) $$F \in \mathcal{B}_1^+ \cap \mathcal{B}_1^- \cap \mathcal{B}_-,$$

property (4.5) of U_- enables us to define

$$G^{\pm, F} : \Sigma \to \tilde{S} \times \mathbb{C} \times \mathbb{R},$$

$$G^{+,F}(\tilde{v}, w) := \iota\{(P_{R,F}^+ \circ P_{1,F})(x_{t_-} + \iota^{-1}(\tilde{v}, w, 0)) - 1\},$$

$$G^{-,F}(\tilde{v}, w) := \iota\{(P_{R,F}^- \circ P_{1,F})(\tilde{x}_{t_-} + \iota^{-1}(\tilde{v}, w, 0)) + 1\}.$$

The definitions of $P_{R,F}^+$ and $P_{R,F}^-$ imply

(4.11)
$$G^{+,F}(\tilde{v}, w) =$$
$$\iota\{\Phi_F(\tau_{R,F}^+[\Phi_F(t_1 - t_-, x_{t_-} + \iota^{-1}(\tilde{v}, w, 0))] + t_1 - t_-,$$
$$x_{t_-} + \iota^{-1}(\tilde{v}, w, 0)) - 1\},$$
$$G^{-,F}(\tilde{v}, w) =$$
$$\iota\{\Phi_F(\tau_{R,F}^-[\Phi_F(t_1 - t_-, \tilde{x}_{t_-} + \iota^{-1}(\tilde{v}, w, 0))] + t_1 - t_-,$$
$$\tilde{x}_{t_-} + \iota^{-1}(\tilde{v}, w, 0)) + 1\}.$$

We write G^+ and G^- for $G^{+,\hat{f}}$ and $G^{-,\hat{f}}$. Note that oddness of f and the uniqueness properties of the maps $\tau^{\pm}_{R,\hat{f}}$ imply that $P^-_{R,\hat{f}}(-\psi) = -P^+_{R,\hat{f}}(\psi)$ for all $\psi \in U_1$, and that $P_{1,\hat{f}}$ is odd. For $(\tilde{v}, w) \in \Sigma$ we obtain, using that $\tilde{x} = -x$,

$$\begin{aligned}
G^-(-\tilde{v}, -w) &= \iota\{(P^-_{R,\hat{f}} \circ P_{1,\hat{f}}(-x_{t_-} + \iota^{-1}(-\tilde{v}, -w, 0)) + 1\} \\
&= \iota\{(P^-_{R,\hat{f}}[-P_{1,\hat{f}}(x_{t_-} + \iota^{-1}(\tilde{v}, w, 0))] + 1\} \\
&= \iota\{-(P^+_{R,\hat{f}} \circ P_{1,\hat{f}})(...) + 1\} \\
&= -\iota\{(P^+_{R,\hat{f}} \circ P_{1,\hat{f}})(...) - 1\} \\
&= -G^+(\tilde{v}, w).
\end{aligned}$$
(4.12)

For $F: C \to \mathbb{R}$ satisfying condition (4.10), we define (analogous to Section 3) the maps $\mathcal{T}_{1,F}: \Sigma \times \{-z_0, z_0\} \to \tilde{S} \times \mathbb{C}_R \times \mathbb{R} \subset X$ by

$$\mathcal{T}_{1,F}(\tilde{y}, w, z_0) := G^{+,F}(\tilde{y}, w), \quad \mathcal{T}_{1,F}(\tilde{y}, w, -z_0) := G^{-,F}(\tilde{y}, w)$$

for $(\tilde{y}, w) \in \Sigma$. Recall that

$$x_{t_-} = -1 + \iota^{-1}(0, 0, z_0), \text{ and } \tilde{x}_{t_-} = 1 + \iota^{-1}(0, 0, -z_0).$$

Assume now that Q and W_Q are as in Proposition 4.6, that $F \in \mathcal{B}_Q$ satisfies (4.10), and that $\psi \in W_Q$ is such that $(\mathcal{T}_{1,F} \circ \mathcal{T}_{0,Q,F})(\iota\psi)$ is defined. Recall the definitions of $\mathcal{T}_{0,Q,F}$ and of $P_{Q,F}$, and set $z := \mathrm{pr}_3 \iota \psi$. Then

$$\mathcal{T}_{0,Q,F}(\iota\psi) = \iota P_{Q,F}(\psi) = (\tilde{v}, w, \mathrm{sign}(z) z_0),$$

with $(\tilde{v}, w) \in \Sigma$. Further,

$$(\mathcal{T}_{1,F} \circ \mathcal{T}_{0,Q,F})(\iota\psi) = \iota\{(P_{R,F} \circ P_{1,F})[\mathrm{sign}(z) \cdot x_{t_-} + \iota^{-1}(\tilde{v}, w, 0)] - \mathrm{sign}(z)\},$$

and the expression in square brackets equals

$$\begin{aligned}
&-\mathrm{sign}(z) + \iota^{-1}(0, 0, \mathrm{sign}(z) z_0) + \iota^{-1}(\tilde{v}, w, 0) \\
&= -\mathrm{sign}(z) + \iota^{-1} \iota P_{Q,F}(\psi) = -\mathrm{sign}(z) + P_{Q,F}(\psi).
\end{aligned}$$

Together, we obtain
(4.13)
$$(\mathcal{T}_{1,F} \circ \mathcal{T}_{0,Q,F})(\iota\psi) = \iota\{(P_{R,F} \circ P_{1,F})(-\mathrm{sign}(z) + P_{Q,F}(\psi)) - \mathrm{sign}(z)\}.$$

The following notation is convenient to express closeness of functionals F to $\hat{f}$: For $F \in BLip(C, \mathbb{R})$ with the property that $F_{|-1+C(l)}$ and $F_{|1+C(l)}$ are BC^1, we define

$$\|F\|_l := \max\{\|F\|_{C^0}, \|F_{|-1+C(l)}\|_{C^1}, \|F_{|1+C(l)}\|_{C^1}\}.$$

4.10. Proposition. *There exists $z_1 \in (0, z_0]$ with the following properties: Assume that $Q \subset U_L$ and W_Q are as in Proposition 4.6, and that for all $\psi \in Q: |\text{pr}_3 \iota \psi| < z_1$.*

a) Then there exist an open neighborhood $\tilde{W}_Q \subset W_Q$ of Q in C and $\theta_Q > 0$ such that if $F \in BC^1(C, \mathbb{R})$ satisfies

(4.10.1) $$\|F - \hat{f}\|_\iota \leq \theta_Q$$

then $F \in \mathcal{B}_1^+ \cap \mathcal{B}_1^- \cap \mathcal{B}_- \cap \mathcal{B}_Q$, and $\mathcal{T}_{1,F} \circ \mathcal{T}_{0,Q,F}$ is defined on $\iota(\tilde{W}_Q)$.

b) For every $\varepsilon > 0$ there exists $\theta(\varepsilon) \in (0, \theta_Q]$ such that if $F \in BC^1(C, R)$ satisfies

(4.10.2) $$\|F - \hat{f}\|_\iota \leq \theta(\varepsilon),$$

and conditions (ii, $\theta(\varepsilon)$) and (4.8.1) of Proposition 4.8, then one has

(4.10.3) $$\|(\mathcal{T}_{1,F} \circ \mathcal{T}_{0,Q,F} - \mathcal{T}_{1,\hat{f}} \circ \mathcal{T}_{0,Q,\hat{f}})|_{\iota(\tilde{W}_Q)}\|_{C^1} \leq \varepsilon.$$

Proof. Choose $z_1 \in (0, z_0]$ such that (with r_- from the definition of U_L) one has for all $z \in [-z_1, z_1] \setminus \{0\}$ the inequality

$$\max\{K \exp(\gamma \tau(z)), \exp(\rho \tau(z))\} < \frac{K r_\Sigma}{2 r_-}.$$

Let $\psi \in U_L$ with $|\text{pr}_3 \iota \psi| < z_1$. Then, with $z := \text{pr}_3 \iota \psi$,

$$\mathcal{T}_0(\iota \psi) = T(\tau(z)) \iota \psi = (\tilde{v}, w, \text{sign}(z) z_0),$$

where $|\tilde{v}| \leq K \exp(\gamma \tau(z)) |\text{pr}_1 \iota \psi|$, and $|w| = \exp(\rho \tau(z)) |\text{pr}_2 \iota \psi|$. Using the definition of U_L and the choice of z_1, we get

(4.10.4) $$|(\tilde{v}, w)| \leq \max\{K \exp(\gamma \tau(z)), \exp(\rho \tau(z))\} |\text{pr}_{1,2} \iota \psi|$$
$$< \frac{K r_\Sigma}{2 r_-} \frac{r_-}{K} = \frac{r_\Sigma}{2}.$$

Let now $Q \subset U_L$ and W_Q be as in Proposition 4.6, and such that for all $\psi \in Q: |\text{pr}_3 \iota \psi| < z_1$.

Proof of a): The set

$$\tilde{W}_Q := W_Q \cap \{\psi \in \mathbb{C} \mid \frac{1}{2} \inf_{\varphi \in Q} |\text{pr}_3 \iota \varphi| < |\text{pr}_3 \iota \psi| < z_1\}$$

is an open neighborhood of Q in C.

(Recall β_Q from Proposition 4.6.) It follows from Proposition 4.6, b) that there exists $\theta'_Q \in (0, \beta_Q]$ such that for all $F \in \mathcal{B}_Q$ with

$\|(F - \hat{f})_{|-1+C(l)}\|_{C^1} \leq \theta'_Q$, one has $\|\iota(P_{Q,F} - P_{Q,\hat{f}})\iota^{-1}{}_{|\iota(W_Q)}\|_{C^0} < r_\Sigma/2$. The definitions of $\mathcal{B}_1^+, \mathcal{B}_1^-$ and $\mathcal{B}_-$ show that there exists $\theta' > 0$ such that $F \in BC^1(C, \mathbb{R})$, $\|F - \hat{f}\|_l \leq \theta'$ implies $F \in \mathcal{B}_1^+ \cap \mathcal{B}_1^- \cap \mathcal{B}_-$. Set $\theta_Q := \min\{\theta'_Q, \theta'\}$. If now $F \in BC^1(C, \mathbb{R})$ satisfies condition (4.10.1) then, using $\theta_Q \leq \theta'_Q \leq \beta_Q$, we see that $F \in \mathcal{B}_Q \cap \mathcal{B}_1^+ \cap \mathcal{B}_1^- \cap \mathcal{B}_-$.

For such F, the definition of $\mathcal{T}_{0,Q,F}$ and (4.10.4) together show that for $\psi \in \tilde{W}_Q$ one has $\mathcal{T}_{0,Q,F}(\iota\psi) = (\tilde{v}, w, z_0 \cdot \text{sign}(\text{pr}_3\iota\psi))$, where $(\tilde{v}, w)$ satisfies $|(\tilde{v}, w)| < r_\Sigma/2 + r_\Sigma/2 = r_\Sigma$, so $(\tilde{v}, w) \in \Sigma$. Hence, $(\mathcal{T}_{1,F} \circ \mathcal{T}_{0,Q,F})(\iota\psi)$ is defined (and equals $G^+(\tilde{v}, w)$ or $G^-(\tilde{v}, w)$).

Proof of b): The maps $P_{Q,\hat{f}|\tilde{W}_Q}$, $P_{1,\hat{f}}$ and $P_{R,\hat{f}}$ are BC^1 maps with uniformly continuous derivatives, and the latter two maps are uniformly continuous (see Propositions 4.6 and 4.7, and the passage before Proposition 4.8). Further, the function $\psi \mapsto \text{sign}\,\text{pr}_3\iota\psi$ is locally constant on $\tilde{W}_Q$, since $\inf_{\psi \in \tilde{W}_Q} |\text{pr}_3\iota\psi| > 0$. We can now apply Proposition 6.6 (from the appendix).

Let $\varepsilon > 0$. It follows from formula (4.13) and Proposition 6.6 that there exists $\delta_1(\varepsilon) > 0$ such that if $F \in BC^1(C, R)$ satisfies (4.10.1) and if
(4.10.5)
$$\max\{\|P_{R,F} - P_{R,\hat{f}}\|_{C^1}, \|P_{1,F} - P_{1,\hat{f}}\|_{C^1}, \|(P_{Q,F} - P_{Q,\hat{f}})_{|\tilde{W}_Q}\|_{C^1}\} \leq \delta_1(\varepsilon)$$

then the estimate (4.10.3) holds. Using Proposition 4.6, b), Proposition 4.7, b), the analogous statement for $P_{R,F}^-$, the definition of $P_{R,F}$, and Proposition 4.8, we obtain $\delta_2(\varepsilon) > 0$ such that if $F \in BC^1(C, \mathbb{R}) \cap \mathcal{B}_-$ satisfies
(4.10.6)
$$\max\{\|(F - \hat{f})_{|1+C(l)}\|_{C^1}, \|(F - \hat{f})_{|-1+C(l)}\|_{C^1}\} \leq \delta_2(\varepsilon),$$

and conditions (i, $\delta_2(\varepsilon)$), (ii, $\delta_2(\varepsilon)$), and (4.8.1) of Proposition 4.8, then (4.10.5) and, consequently, (4.10.3) holds.

Set $\theta(\varepsilon) := \min\{\delta_2(\varepsilon), \theta_Q\}$. If now $F \in BC^1(C, \mathbb{R})$ is as in assertion b), then F satisfies (4.10.1) and thus, in particular, $F \in \mathcal{B}_-$. Further, F satisfies conditions (4.8.1), (i, $\delta_2(\varepsilon)$) and (ii, $\delta_2(\varepsilon)$) of Proposition 4.8, together with the first part of (4.10.6). Thus the conclusion (4.10.3) holds. □

With the last proposition we have provided the perturbation statement that later allows to extend results for one particular delay equation to nearby equations.

Our next aim is to verify that orbits of the maps $\mathcal{T}_{1,F} \circ \mathcal{T}_{0,Q,F}$ define solutions of equation (F), and to describe the oscillation properties of these solutions.

4.11. Remark. *Let Q and $\mathcal{B}_Q$ and W_Q be as in Proposition 4.6, and assume $\psi \in W_Q$ and*

(4.11.1) $\quad F \in \mathcal{B}_1^+ \cap \mathcal{B}_1^- \cap \mathcal{B}_- \cap \mathcal{B}_Q, \quad F(\varphi + 2) = F(\varphi)\ (\varphi \in C).$

Set $z := \mathrm{pr}_3 \iota\psi$, and assume that $\psi' := \iota^{-1}(\mathcal{T}_{1,F} \circ \mathcal{T}_{0,Q,F})(\iota\psi)$ is defined (i.e., $\mathrm{pr}_{1,2}\mathcal{T}_{0,Q,F}(\iota\psi) \in \Sigma$)). Then the number

$$\tau^*(\psi) :=$$
$$\begin{cases} \tau_{R,F}^+(\Phi_F[t_1 - t_-, \Phi_F(\tau_{Q,F}(\psi), -1+\psi)]) + t_1 - t_- + \tau_{Q,F}(\psi) & \text{if } z > 0, \\ \tau_{R,F}^-(\Phi_F[t_1 - t_-, \Phi_F(\tau_{Q,F}(\psi), 1+\psi)]) + t_1 - t_- + \tau_{Q,F}(\psi) & \text{if } z < 0 \end{cases}$$

is well-defined, and one has

(4.11.2) $\qquad \psi' = \Phi_F(\tau^*(\psi), -\mathrm{sign}\,(z) + \psi) - \mathrm{sign}\,(z).$

Proof. Recall formula (4.13). From the definition of $P_{Q,F}$, we have in case $z > 0$

$$-\mathrm{sign}\,(z) + P_{Q,F}(\psi) = -1 + P_{Q,F}(\psi) = \Phi_F(\tau_{Q,F}(\psi), -1+\psi).$$

Note that the periodicity condition on F implies that
$\Phi_F(t, \varphi) + 2 = \Phi_F(t, \varphi + 2)$ ($\varphi \in C$, $t \geq 0$). In case $z < 0$, we thus have

$$\begin{aligned} -\mathrm{sign}\,(z) + P_{Q,F}(\psi) &= 1 + P_{Q,F}(\psi) = 1 + \Phi_F(\tau_{Q,F}(\psi), -1+\psi) + 1 \\ &= 2 + \Phi_F(\tau_{Q,F}(\psi), -1+\psi) \\ &= \Phi_F(\tau_{Q,F}(\psi), 1+\psi). \end{aligned}$$

Now the definitions of τ^*, $P_{R,F}$ and $P_{1,F}$, and the fact that $(\mathcal{T}_{1,F} \circ \mathcal{T}_{0,Q,F})(\iota\psi)$ is defined, together with the semigroup property of Φ_F imply that, in both cases, $\tau^*(\psi)$ is well-defined and

$$(P_{R,F} \circ P_{1,F})(-\mathrm{sign}\,(z) + P_{Q,F}(\psi)) = \Phi_F(\tau^*(\psi), -\mathrm{sign}\,(z) + \psi).$$

The assertion now follows from (4.13) by application of ι^{-1} from the left. $\square$

We can now show that orbits of the maps $\mathcal{T}_{1,F} \circ \mathcal{T}_{0,Q,F}$ define solutions of equation (F) which follow level sequences in the sense of Definition 4.2.

4.12. Proposition. *Let $Q \subset U_L$ be as in Proposition 4.6, and assume that, with $\mathcal{K}$ from Proposition 4.4, $Q \subset \mathcal{K}$. Assume that F satisfies condition (4.11.1). With $\mathbb{M} := \mathbb{N}_0$ or $\mathbb{M} = \mathbb{Z}$, let $(\tilde{x}_j, w_j, z_j)_{j \in \mathbb{M}}$ be a trajectory of $\mathcal{T}_{1,F} \circ \mathcal{T}_{0,Q,F}$ in $\iota(Q)$.*
Define the level sequence $\mathbf{l} = (l_j)_{j \in \mathbb{M}}$ by

$$l_0 := -1, \quad l_{j+1} := l_j + 2\,\mathrm{sign}\,(z_j) \quad (j \in \mathbb{M}).$$

(Then $\mathbf{l} \in \mathcal{L}^+$ if $\mathbb{M} = \mathbb{N}_0$ and $\mathbf{l} \in \mathcal{L}$ if $\mathbb{M} = \mathbb{Z}$.) Set $\psi_j := \iota^{-1}(\tilde{x}_j, w_j, z_j)$ and $\varphi_j := l_j + \psi_j$ ($j \in \mathbb{M}$), and define $(\tau_j)_{(j \in \mathbb{M})}$ by

$$\tau_0 = 0, \quad \tau_{j+1} = \tau_j + \tau^*(\psi_j) \quad (j \in \mathbb{M}).$$

There exists a unique solution $y : [-1, \infty) \to \mathbb{R}$ (in case $\mathbb{M} = \mathbb{N}_0$), respectively $y : \mathbb{R} \to \mathbb{R}$ (in case $\mathbb{M} = \mathbb{Z}$) with $y_{\tau_j} = \varphi_j$ $(j \in \mathbb{M})$, and y follows the level sequence $\mathbf{l}$.

Proof. It follows from $(\tilde{x}_{j+1}, w_{j+1}, z_{j+1}) = (\mathcal{T}_{1,F} \circ \mathcal{T}_{0,Q,F})(\tilde{x}_j, w_j, z_j)$ $(j \in \mathbb{M})$ and Remark 4.11 that $\tau^*(\psi_j)$ is defined.
(4.12.2) *Claim:* $\varphi_{j+1} = \Phi_F(\tau^*(\psi_j), \varphi_j)$ $(j \in \mathbb{M})$. *Proof:*

$$\begin{aligned}
\varphi_{j+1} &= l_{j+1} + \iota^{-1}(\tilde{x}_{j+1}, w_{j+1}, z_{j+1}) \\
&= l_j + 2\operatorname{sign}(z_j) + \iota^{-1}(\mathcal{T}_{1,F} \circ \mathcal{T}_{0,Q,F})(\tilde{x}_j, w_j, z_j) \\
&= l_j + 2\operatorname{sign}(z_j) + \iota^{-1}(\mathcal{T}_{1,F} \circ \mathcal{T}_{0,Q,F})(\iota \psi_j) \\
&= l_j + 2\operatorname{sign}(z_j) + \Phi_F(\tau^*(\psi_j), -\operatorname{sign}(z_j) + \psi_j) - \operatorname{sign}(z_j) \\
&\quad \text{(from (4.11.2))} \\
&= l_j + \operatorname{sign}(z_j) + \Phi_F(\tau^*(\psi_j), -\operatorname{sign}(z_j) + \psi_j) \\
&= \Phi_F(\tau^*(\psi_j), l_j + \psi_j) = \Phi_F(\tau^*(\psi_j), \varphi_j),
\end{aligned}$$

where we used periodicity of $\Phi_F(\tau^*(\psi_j), \cdot)$ and the fact that $l_j + \operatorname{sign}(z_j) \in 2\mathbb{Z}$. (Claim (4.12.2) is proved.)

Note that the definition of τ^* and $t_1 - t_- > 1$ imply that $\tau_{j+1} - \tau_j > 1$ $(j \in \mathbb{M})$. It follows from the above claim and the definition of $(\tau_j)_{j \in \mathbb{M}}$ that there exists a (unique) solution $y : [-1, \infty) \to \mathbb{R}$, respectively $y : \mathbb{R} \to \mathbb{R}$, of equation (F) with $y_{\tau_j} = \varphi_j$ $(j \in \mathbb{M})$. Uniqueness of this solution is clear.
(4.12.3) *Claim:* y wanders from l_j to l_{j+1} on $[\tau_j - 1, \tau_{j+1}]$ $(j \in \mathbb{M})$.
Proof: Let $j \in \mathbb{M}$. We have $y_{\tau_j} = \varphi_j = l_j + \psi_j$. Since $\psi_j \in Q \subset \mathcal{K}$, Proposition 4.4 shows that $\psi_j([-1, 0]) \cap \mathbb{Z} \supset \{0\}$. Further, since $Q \subset U_L$, we have $|\psi_j| \leq l < 1/2$, so together we obtain

(4.12.4) $\qquad \psi_j([-1, 0]) \cap \mathbb{Z} = \{0\}$.

Let $\tilde{y} : [-1, \infty) \to \mathbb{R}$ denote the solution of equation (F) with $\tilde{y}_0 = -\operatorname{sign}(z_j) + \psi_j$. Periodicity of F, and $l_j \in 2\mathbb{Z} + 1$, and $y_{\tau_j} = l_j + \psi_j$ imply

(4.12.5) $\qquad \forall t \geq -1 : \quad y(\tau_j + t) = l_j + \operatorname{sign}(z_j) + \tilde{y}(t)$.

Let $\tilde{\tilde{y}} : [-1, \infty) \to \mathbb{R}$ denote the solution of equation (F) with $\tilde{\tilde{y}}_0 = -1 + \psi_j$. In case $z_j > 0$ we have $\tilde{\tilde{y}} = \tilde{y}$, and in case $z_j < 0$, periodicity of F implies $\tilde{y} = \tilde{\tilde{y}} + 2$. It follows from Proposition 4.6, c) that

$$\tilde{\tilde{y}}([-1, \tau_{Q,F}(\psi_j)]) \cap \mathbb{Z} \subset \{-1\},$$

and now (4.12.4) shows that $\tilde{\tilde{y}}([-1, \tau_{Q,F}(\psi_j)]) \cap \mathbb{Z} = \{-1\}$. Hence we have in both cases ($z_j > 0$ and $z_j < 0$)

(4.12.6) $\qquad \tilde{y}([-1, \tau_{Q,F}(\psi_j)]) \cap \mathbb{Z} = \{-\operatorname{sign}(z_j)\}$.

The fact that $(\mathcal{T}_{1,F} \circ \mathcal{T}_{0,Q,F})(\iota\psi_j)$ is defined implies that in case $z_j > 0$ one has
$$\tilde{y}_{\tau_{Q,F}(\psi_j)} = \Phi_F(\tau_{Q,F}(\psi_j), -1 + \psi_j) \in \Sigma_C \subset U_-,$$
and in case $z_j < 0$, one has $\tilde{y}_{\tau_{Q,F}(\psi_j)} = \Phi_F(\tau_{Q,F}(\psi_j), 1 + \psi_j) \in -\Sigma_C \subset (-U_-)$. We obtain from the first property in (4.6) that in both cases

(4.12.7) $\quad \tilde{y}([\tau_{Q,F}(\psi_j) - 1, \tau_{Q,F}(\psi_j) + t_1 - t_- - 1]) \cap \mathbb{Z} = \{0\}.$

From the last line and from (4.12.6) we conclude

(4.12.8) $\quad \tilde{y}([-1, \tau_{Q,F}(\psi_j) - 1]) \cap \mathbb{Z} = \{-\text{sign}(z_j)\}.$

Further, we see from (4.6) that
$$\tilde{y}([\tau_{Q,F}(\psi_j) + t_1 - t_- - 1, \tau_{Q,F}(\psi_j) + t_1 - t_-]) \cap \mathbb{Z} = \{\text{sign}(z_j)\}.$$

From (4.5), we have $\tilde{y}_{\tau_{Q,F}(\psi_j)+t_1-t_-} \in \text{sign}(z_j) \cdot U_1$. Recall that $\pm U_1 \subset U_1^{\pm}$. Now Proposition 4.7, c), and the analogous properties for $\tau_{R,F}^-$ and U_1^- show that with
$$\theta := \begin{cases} \tau_{R,F}^+(\tilde{y}_{\tau_{Q,F}(\psi_j)+t_1-t_-}) & \text{in case } z_j > 0 \\ \tau_{R,F}^-(\tilde{y}_{\tau_{Q,F}(\psi_j)+t_1-t_-}) & \text{in case } z_j < 0, \end{cases}$$
one has $\tilde{y}([\tau_{Q,F}(\psi_j) + t_1 - t_- - 1, \tau_{Q,F}(\psi_j) + t_1 - t_- + \theta]) \cap \mathbb{Z} = \{\text{sign}(z_j)\}$. The right endpoint of the interval in the last formula equals in both cases $\tau^*(\psi_j)$, so we have

(4.12.9) $\quad \tilde{y}([\tau_{Q,F}(\psi_j) + t_1 - t_- - 1, \tau^*(\psi_j)]) \cap \mathbb{Z} = \{\text{sign}(z_j)\}.$

Properties (4.12.8), (4.12.7) and (4.12.9) together show that $\tilde{y}$ wanders from $-\text{sign}(z_j)$ to $\text{sign}(z_j)$ on the interval $[-1, \tau^*(\psi_j)]$. With (4.12.5), it follows that y wanders from l_j to $l_j + 2\text{sign}(z_j)$ on the interval $[\tau_j - 1, \tau_j + \tau^*(\psi_j)]$. Since $\tau_j + \tau^*(\psi_j) = \tau_{j+1}$ and $l_{j+1} = l_j + 2\text{sign}(z_j)$, Claim (4.12.3) is proved.

Since $\tau_{j+1} > \tau_j + 1$ $(j \in \mathbb{M})$ and since τ^* is bounded, the intervals $[\tau_j - 1, \tau_{j+1}]$ $(j \in \mathbb{M})$ have the properties required in part c) of Definition 4.1. Hence claim (4.12.3) implies that y follows the level sequence l. $\square$

Of course, we want to apply the results of Section 3 in order to actually obtain trajectories of the maps $\mathcal{T}_{1,F} \circ \mathcal{T}_{0,Q,F}$, the existence of which was *assumed* in Proposition 4.12. From Proposition 4.3 and condition (f3), we know already that the semigroup T used in the definition of $\mathcal{T}_0$ satisfies conditions (T1)-(T4) of Section 3. Since $P_{R,\hat{f}}^+$, $P_{R,\hat{f}}^-$, and $P_{1,\hat{f}}$ are BC^1, we know that G^+ and G^- are BC^1.

4.13. Proposition. *The maps $G^\pm$ satisfy conditions* (G1) *and* (G2) *of Section 3.*

Proof. Since $R = |\text{pr}_2\iota(x_{t_+} - 1)| = |\text{pr}_2\iota[\Phi_f(t_+ - t_1, x_{t_1}) - 1]|$, the property $t_+ - t_1 \in I_R^+$ and the uniqueness property of $\tau_{R,\hat{f}}^+$ imply that $\tau_{R,\hat{f}}^+(x_{t_1}) = t_+ - t_1$, and hence $P_{R,\hat{f}}^+(x_{t_1}) = x_{t_+}$. Now

$$G^+(0,0) = \iota[(P_{R,\hat{f}}^+ \circ P_{1,\hat{f}})(x_{t_-}) - 1] = \iota[P_{R,\hat{f}}^+(x_{t_1}) - 1] = \iota(x_{t_+} - 1).$$

Since $x_t - 1 \in \tilde{S} \oplus S_2$ for $t \geq t_1$, we have $\text{pr}_3 G^+(0,0) = 0$. From (4.12) we infer that also $\text{pr}_3 G^-(0,0) = 0$. (Property (G1) is proved.)

Proof of (G2): We have

$$DG^+(0,0) = \iota \circ DP_{R,\hat{f}}(x_{t_1}) \circ DP_{1,\hat{f}}(x_{t_-}) \circ \iota^{-1}(\cdot,\cdot,0).$$

The tangent space to the set $\{\psi \in C \mid |\text{pr}_2\iota(\psi - 1)| = R\}$ at x_{t_+} is given by

$$H := \iota^{-1}\{\tilde{S} \times (\mathbb{R} \cdot \exp(i\pi/2)\text{pr}_2\iota(x_{t_+} - 1)) \times \mathbb{R}\}.$$

Note that

$$\begin{aligned}
\text{pr}_2 \iota \dot{x}_{t_+} &= \frac{d}{dt}[t \mapsto \text{pr}_2\iota(x_t - 1)]\big|_{t=t_+} \\
&= \frac{d}{dt}[t \mapsto \exp(\mu(t-t_1))\text{pr}_2\iota(x_{t_1} - 1)]\big|_{t=t_+} \\
&= \mu \text{pr}_2\iota(x_{t_+} - 1),
\end{aligned}$$

and hence $\dot{x}_{t_+} \notin H$, since $\mu \notin \mathbb{R} \cdot \exp(i\pi/2)$.

Let $\text{pr}_H \in L_c(C,C)$ denote the projection onto H parallel to $\dot{x}_{t_+}$. There exists an $\mathbb{R}$-linear functional $\xi : C \to \mathbb{R}$ with $\xi(\dot{x}_{t_+}) = 1$ such that

$$\text{pr}_H \psi = \psi - \xi(\psi) \cdot \dot{x}_{t_+} \quad (\psi \in C).$$

Since $\dot{x}_{t_+} \in \tilde{S} \oplus S_2$, we conclude that

$$\text{pr}_3 \circ \iota \circ \text{pr}_H = \text{pr}_3 \circ \iota.$$

Since $\hat{f}$ is C^1 on $1 + C(l)$, it follows from property (x3) and from Proposition 3.2, p. 370 of [Diekmann et al.] that

$$DP_{R,\hat{f}}(x_{t_1}) = \text{pr}_H \circ D_2\Phi_f(t_+ - t_1, x_{t_1}).$$

Further, properties (x3) and (f1), (f2) imply that

$$D_2\Phi_f(t_+ - t_1, x_{t_1}) = T^\alpha(t_+ - t_1).$$

Together we obtain

$$\begin{aligned}
\mathrm{pr}_3 DG^+(0,0) &= \mathrm{pr}_3 \iota \mathrm{pr}_H D_2 \Phi_f(t_+ - t_1, x_{t_1}) DP_{1,\hat{f}}(x_{t_-}) \iota^{-1}(\cdot, \cdot, 0) \\
&= \mathrm{pr}_3 \iota D_2 \Phi_f(t_+ - t_1, x_{t_1}) DP_{1,\hat{f}}(x_{t_-}) \iota^{-1}(\cdot, \cdot, 0) \\
&= \mathrm{pr}_3 \iota T^\alpha(t_+ - t_1) DP_{1,\hat{f}}(x_{t_-}) \iota^{-1}(\cdot, \cdot, 0) \\
&= \mathrm{pr}_3 \iota T^\alpha(t_+ - t_1) \iota^{-1} \iota DP_{1,\hat{f}}(x_{t_-}) \iota^{-1}(\cdot, \cdot, 0) \\
&= \exp(\lambda(t_+ - t_1)) \mathrm{pr}_3 \iota DP_{1,\hat{f}}(x_{t_-}) \iota^{-1}(\cdot, \cdot, 0).
\end{aligned}$$

Since $\iota^{-1}(\{0\} \times \mathbb{C} \times \{0\}) = S_2$, property (G2) for G^+ is proved if we show

(4.13.1) $$\mathrm{pr}_3 \iota DP_{1,\hat{f}}(x_{t_-})\big|_{S_2} \neq 0.$$

Recall now the solution v of the variational equation (f, x) with $v(t) = \exp(\mu t)$ ($t \leq 0$). The functions $a := \mathrm{Re}(v)$, $b := \mathrm{Im}(v)$ are also solutions of (f, x). The definition of S_2 implies that, since $a(t) = e^{\mu t}/2 + e^{\bar{\mu} t}/2$, $b(t) = e^{\mu t}/2i + e^{\bar{\mu} t}/(-2i)$ for $t \leq 0$, we have

(4.13.2) $$a_{t_-}, b_{t_-} \in S_2.$$

For $\chi \in C$, $\mathrm{pr}_3 \iota \chi$ is given by

$$\frac{1}{1+\lambda}[\chi(0) + \lambda \int_{-1}^0 \exp(-\lambda s) \chi(s) ds]$$

(see Lemma 6.8, b)). Hence condition (v) is equivalent to saying that

$$\mathrm{pr}_3 \iota a_{t_1} + i \cdot \mathrm{pr}_3 \iota b_{t_1} \neq 0,$$

so that at least one of the numbers $\mathrm{pr}_3 \iota a_{t_1}$, $\mathrm{pr}_3 \iota b_{t_1}$ is nonzero.

From the construction of $P_{1,\hat{f}}$ and U_- (see the passage before Proposition 4.8), and from the fact that a, b are solutions of the variational equation (f, x), we infer that

$$DP_{1,\hat{f}}(x_{t_-}) a_{t_-} = a_{t_1}, \quad DP_{1,\hat{f}}(x_{t_-}) b_{t_-} = b_{t_1},$$

so we have $\mathrm{pr}_3 \iota DP_{1,\hat{f}}(x_{t_-}) a_{t_-} \neq 0$ or $\mathrm{pr}_3 \iota DP_{1,\hat{f}}(x_{t_-}) b_{t_-} \neq 0$. In view of (4.13.2), property (4.13.1) is proved. Property (G2) for the map G^- now follows from (4.12). □

Recall the notion of symbolic dynamics with respect to a set of level sequences (from Definition 4.2), and the definitions of the sets $\mathcal{L}$, $\mathcal{L}^+$ (before Proposition 4.4). In the next theorem, the main result of this section, we combine the preceding statements to derive the existence of solutions of equation (f) which follow level sequences from $\mathcal{L}^+$ and from $\mathcal{L}$. In Theorem 4.15 below, we partially extend the result to nearby equations which do not necessarily satisfy linearity conditions.

4.14 Theorem. *Assume that conditions* (f1)–(f3) *on the nonlinearity* f, *conditions* (x1)–(x4) *on the heteroclinic solution* x, *and condition* (v) *on the variational equation* (f, x) *all hold.*

a) *Then there exist* $k^* \in \mathbb{N}$ *and countably many closed sets* R^k $(k \geq k^*)$ *such that equation* (f) *has symbolic dynamics with respect to* $\mathcal{L}$ *on each set* $-1 + R^k$ $(k \geq k^*)$.

b) *The sets* R^k *have the following form: Let* $C = \tilde{S} \oplus S_2 \oplus U$ *be the decomposition according to Proposition 4.3. There exist closed subsets* I_{-1}, I_1 *of* $\tilde{S} \oplus S_2$ *and closed intervals* $J_l^k = [a_l^k, b_l^k] \subset \mathbb{R}_+$ $(l = -1, 1)$ *with*

$$J_l^{k+1} = \exp(-2\pi\lambda/\omega) \cdot J_l^k \quad (l = -1, 1, k \geq k^*)$$

such that, with $\tilde{J}_l^k := J_l^k \cdot \exp(\lambda \cdot) \subset U$, *one has*

$$R^k = [I_1 + (\tilde{J}_1^k \cup (-\tilde{J}_1^k))] \cup [I_{-1} + (\tilde{J}_{-1}^k \cup (-\tilde{J}_{-1}^k))] \quad (l = -1, 1, k \geq k^*).$$

If $I_1 \neq I_{-1}$ *then* $\mathrm{dist}(I_1, I_{-1}) > 0$.

c) *Let a sequence* $\mathbf{l} = (l_0, l_1, ...) \in \mathcal{L}^+$ *be given. For every* $k \geq k^*$ *there exists a Lipschitz continuous function* $\varphi_{k,\mathbf{l}} : I_1 \cup I_{-1} \to U$ *with* $\varphi_{k,\mathbf{l}}(I_l) \subset \mathrm{sign}\,(l_1 - l_0)\tilde{J}_l^k$ $(l = -1, 1)$ *such that for all* $\psi \in \mathrm{graph}\,\varphi_{k,\mathbf{l}}$, *the solution* $y^{-1+\psi}$ *of equation* (f) *with* $y_0^{-1+\psi} = -1 + \psi$ *follows the level sequence* $\mathbf{l}$.

Proof. We know that the semigroup T used in the definition of $\mathcal{T}_0$ satisfies conditions (T1)-(T3), and, from Proposition 4.13, that the maps $G^\pm$ satisfy conditions (G1) and (G2) of Section 3. (We first define $k^* \in \mathbb{N}$ and sets R^k as in assertion b).) Recall $\beta \in (0, 1]$ from Proposition 4.4. Set $\tilde{x}_\pm := \mathrm{pr}_1 G^\pm(0, 0) = \pm \mathrm{pr}_1 \iota(x_{t_+} - 1)$; then property (t^+, i) shows that

$$|\tilde{x}_\pm| < \frac{\beta}{2}|\mathrm{pr}_2 \iota(x_{t_+} - 1)| = \beta R/2.$$

Choose $\tilde{\delta} \in (0, \pi/2)$ and apply Lemma 3.2 with $\tilde{r} := \beta R/2$. One obtains $k_0 \in \mathbb{N}$ and sets

$$\Delta^{(k)} = \bigcup_{l,m=+,-} I_l \times J_{l,m}^{(k)} \quad (k \geq k_0)$$

as in assertion A1) of Lemma 3.2, with $I_l \subset \tilde{S} \times \mathbb{R}$, $J_{l,m}^{(k)} \subset \mathbb{R}$, and the property $|\tilde{x} - \tilde{x}_l| \leq \tilde{r} = \beta R/2$ for all $(\tilde{x}, \varphi) \in I_l$ $(l, m = +, -)$. Hence we have

(4.14.1) $\quad |\tilde{x}| \leq \beta R/2 + \beta R/2 = \beta R$ for all $(\tilde{x}, \varphi) \in I_l$ $(l = +, -)$.

It follows from (4.12) that $DG^+(0, 0) = DG^-(0, 0)$, and therefore we have $\theta_{0,+} = \theta_{0,-}$ (compare (3.2.1)). Hence, we see from the definition of the numbers $a_{l,m}^{(k)}, b_{l,m}^{(k)}$ $(l, m = +, -)$ in the proof of Lemma 3.2 that

(4.14.2) $\quad J_{l,+}^{(k)} = -J_{l,-}^{(k)} \quad (l = +, -, k \geq k_0)$.

Set $J_1^k := J_{+,+}^{(k)}$, $J_{-1}^k := J_{-,+}^{(k)}$ for $k \geq k_0$. We see from A1) of Lemma 3.2 that

(4.14.3) $$J_l^{k+1} = \exp(-2\pi\lambda/\omega)J_l^k \quad (l = -1, 1, \ k \geq k_0).$$

Set $\mathcal{T}_1 := \mathcal{T}_{1,\hat{f}}$. With κ and $\chi = \kappa^{-1} \circ \mathcal{T}_1 \circ \mathcal{T}_0 \circ \kappa$ as in A3) and A4) of Lemma 3.2, we know that χ is C^1 on a set D which is an open neighborhood of $\Delta^{(k)}$ ($k \geq k_0$) and satisfies the conditions of Theorem 2.4. (Note that $J_{1,1}, J_{-1,1}, J_{1,-1}, J_{-1,-1}$ in Theorem 2.4 correspond to $J_{+,+}^{(k)}, J_{-,+}^{(k)}, J_{+,-}^{(k)}, J_{-,-}^{(k)}$ here.) Set

$$I_1 := \iota^{-1}\kappa(I_+ \times \{0\}), \quad I_{-1} := \iota^{-1}\kappa(I_- \times \{0\}).$$

Then $I_1 = \{\psi \in C \mid \exists (\tilde{x}, \varphi) \in I_+ : \iota\psi = (\tilde{x}, R\exp(i\varphi), 0)\}$, and the analogous relation holds for I_{-1} and I_-. Obviously I_1 and I_{-1} are closed subsets of $\tilde{S} \oplus S_2$. It follows from the construction of $I_\pm$ in Lemma 3.2 that $\text{dist}(I_1, I_{-1}) > 0$ if $I_1 \neq I_{-1}$ (compare (3.2.4)b)).

With R^k defined as in part b) of the present theorem, we have for $k \geq k_0$

$$R^k = \iota^{-1}\kappa[I_+ \times (J_1^k \cup (-J_1^k)) \cup I_- \times (J_{-1}^k \cup (-J_{-1}^k))]$$
$$= \iota^{-1}\kappa[I_+ \times (J_{+,+}^{(k)} \cup J_{+,-}^{(k)}) \cup I_- \times (J_{-,+}^{(k)} \cup J_{-,-}^{(k)})]$$
$$= \iota^{-1}\kappa\Delta^{(k)}.$$

For $k \geq k_0$ and $\psi \in R^k$, one has $|\text{pr}_2\iota\psi| = R$ and, from (4.14.1),

$$|\text{pr}_1\iota\psi| \leq \beta R = \beta|\text{pr}_2\iota\psi|.$$

One sees from (4.14.3) that there exists $k^* \geq k_0$ such that

$$\forall k \geq k^*, l = \pm 1: \ J_l^k \subset (0, \beta R),$$

and hence

$$\forall k \geq k^* \ \forall \psi \in R^k: \quad |\text{pr}_3\iota\psi| \leq \beta R = \beta|\text{pr}_2\iota\psi|.$$

From property (t_+,iii) and the definition of U_L, we see that $R^k \subset U_L$ for $k \geq k^*$. The definition of $\mathcal{K}$ (Proposition 4.4) and the above estimates on $|\text{pr}_3\iota\psi|, |\text{pr}_1\iota\psi|$ now show that we have

(4.14.4) $$\forall k \geq k^*: \quad R^k \subset \mathcal{K} \cap U_L.$$

Proof of c): Let $\mathbf{l} = (l_0, l_1, ...) \in \mathcal{L}^+$. Define $\mathbf{s} \in \mathcal{S}^+$ by $s_j := \text{sign}(l_{j+1} - l_j)$, $(j \in \mathbb{N}_0)$. Let $k \geq k^*$. Applying Theorem 2.4 with $\Delta := \Delta^{(k)}$, we obtain a Lipschitz continuous function

$$\varphi_\mathbf{s} : I_+ \cup I_- \to \mathbb{R}$$

with the properties a), b), c) stated in Theorem 2.4. Define $\varphi_{k,1} : I_1 \cup I_{-1} \to \mathbb{R} \cdot \exp(\lambda \cdot) \, (= U)$ by

$$\varphi_{k,1}(\psi) := \varphi_{\mathbf{s}}[\mathrm{pr}_{1,2}\kappa^{-1}(\mathrm{pr}_1\iota\psi, \mathrm{pr}_2\iota\psi, 0)] \cdot \exp(\lambda \cdot)$$

(where $\mathrm{pr}_{1,2}(\tilde{x}, \theta, z) := (\tilde{x}, \theta)$). The definition of I_1 and I_{-1} shows that $\varphi_{k,1}$ is well-defined. Recall from A3) of Lemma 3.2 that κ^{-1} is Lipschitz continuous; hence $\varphi_{k,1}$ is Lipschitz continuous. Since sign $(l_1 - l_0) = s_0$, property a) from Theorem 2.4, property (4.14.2), and the definition of J_m^k show that $\varphi_{\mathbf{s}}$ maps $I_{\pm}$ into the set sign $(l_1 - l_0) \cdot \tilde{J}_{\pm 1}^k$, and hence $\varphi_{k,1}$ maps I_m into the set sign $(l_1 - l_0) \cdot \tilde{J}_m^k$ $(m = -1, 1)$.

Let now $\psi \in \mathrm{graph}\,\varphi_{k,1}$; then $\mathrm{pr}_3\iota\psi = \mathrm{pr}_3\iota\varphi_{k,1}(\psi)$. Setting $(\tilde{x}, \theta, z) := \kappa^{-1}\iota\psi$, one has $(\tilde{x}, \theta) \in I_+ \cup I_-$, and

$$z = \mathrm{pr}_3\iota\psi = \varphi_{\mathbf{s}}[\mathrm{pr}_{1,2}\kappa^{-1}(\mathrm{pr}_1\iota\psi, \mathrm{pr}_2\iota\psi, 0)] = \varphi_{\mathbf{s}}(\tilde{x}, \theta),$$

so $(\tilde{x}, \theta, z) \in \mathrm{graph}\,\varphi_{\mathbf{s}}$. From part c) of Theorem 2.4, we know that $(\tilde{x}, \theta, z) \in \mathrm{inv}^+(\chi, \Delta^{(k)})$, and sign $\mathrm{pr}_3\chi^j(\tilde{x}, \theta, z) = s_j$ $(j \in \mathbb{N}_0)$. Since $\chi = \kappa^{-1} \circ \mathcal{T}_1 \circ \mathcal{T}_0 \circ \kappa$ on $\Delta^{(k)}$, it follows from Remark 2.3 that

$$\kappa(\tilde{x}, \theta, z) = (\tilde{x}, R\exp(i\theta), z) \in \mathrm{inv}^+(\mathcal{T}_1 \circ \mathcal{T}_0, \kappa(\Delta^{(k)})),$$

and that

$$\mathrm{sign}\,\mathrm{pr}_3(\mathcal{T}_1 \circ \mathcal{T}_0)^j(\tilde{x}, R\exp(i\theta), z) = s_j \quad (j \in \mathbb{N}_0).$$

We apply Proposition 4.12 with $Q := R^k \subset \mathcal{K} \cap U_L$ and with $F := \hat{f}$, $\mathbb{M} := \mathbb{N}_0$ and with the trajectory defined by

$$(\tilde{x}_j, w_j, z_j) := (\mathcal{T}_1 \circ \mathcal{T}_0)^j(\tilde{x}, R\exp(i\theta), z) \quad (j \in \mathbb{N}_0).$$

Let us denote the level sequence 'l' (from $\mathcal{L}^+$), which was constructed from the sequence $(s_j)_{j\in\mathbb{N}_0} = (\mathrm{sign}\,(z_j))_{j\in\mathbb{N}_0}$ in Proposition 4.12, by $\tilde{\mathbf{l}} = (\tilde{l}_0, \tilde{l}_1, ...)$, for the moment. Then

$$\tilde{l}_0 = -1, \quad \tilde{l}_{j+1} = \tilde{l}_j + 2\mathrm{sign}\,(z_j) = \tilde{l}_j + 2s_j \, (j \in \mathbb{N}_0).$$

In view of the definition of $\mathbf{s}$, and the definition of $\mathcal{L}^+$, our given level sequence $\mathbf{l}$ satisfies

$$l_0 = -1, \quad l_{j+1} = l_j + 2\mathrm{sign}\,(l_{j+1} - l_j) = l_j + 2s_j \, (j \in \mathbb{N}_0).$$

Hence it is obvious that $\tilde{\mathbf{l}} = \mathbf{l}$. Further, ψ_0 from Proposition 4.12 satisfies

$$\psi_0 = \iota^{-1}(\tilde{x}_0, w_0, z_0) = \iota^{-1}(\tilde{x}, R\exp(i\theta), z) = \iota^{-1}\kappa(\tilde{x}, \theta, z) = \psi.$$

Since $l_0 = -1$, we see from Proposition 4.12 that the solution $y^{-1+\psi}$: $[-1, \infty) \to \mathbb{R}$ of equation (f) with $y_0 = -1 + \psi$ follows the level sequence **l**. Part c) is proved.

Proof of a): Let $k \geq k^*$. The fact that $t_1 - t_- > 1$ implies that $\Phi_f(t_1 - t_-, \cdot)$ maps bounded sets to relatively compact sets. The definitions of $\mathcal{T}_1 \circ \mathcal{T}_0$ and of $\mathcal{X}$ show that the set $\operatorname{clos}(\mathcal{X}(\Delta^{(k)}))$ is compact. From part c) of Theorem 2.4, we know that $\mathcal{X}$ has symbolic dynamics with respect to $\mathcal{S}^+$ on $\Delta^{(k)}$. Hence, Proposition 2.2 yields that $\mathcal{X}$ has symbolic dynamics with respect to $\mathcal{S}$ on $\Delta^{(k)}$. Let $\mathbf{l} = (l_j)_{j \in \mathbb{Z}} \in \mathcal{L}$ be given, and define $s_j := \operatorname{sign}(l_{j+1} - l_j)$ $(j \in \mathbb{Z})$. There exists $((\tilde{x}_j, \theta_j, z_j))_{j \in \mathbb{Z}} \in \operatorname{traj}(\mathcal{X}, \Delta^{(k)})$ with $\operatorname{sign}(z_j) = s_j$ $(j \in \mathbb{Z})$. Note that $\kappa(\Delta^{(k)}) = \iota(R^k)$. In view of Remark 2.3, we have

$$((\tilde{x}_j, R\exp(i\theta_j), z_j))_{j \in \mathbb{Z}} \in \operatorname{traj}(\mathcal{T}_1 \circ \mathcal{T}_0, \kappa(\Delta^{(k)})) = \operatorname{traj}(\mathcal{T}_1 \circ \mathcal{T}_0, \iota(R^k)),$$

and, again, the level sequence 'l' constructed in Proposition 4.12 coincides with our given level sequence **l**. Analogously to the proof of c), we obtain the following result from Proposition 4.12: With $\psi_j := \iota^{-1}(\tilde{x}_j, R\exp(i\theta_j), z_j) \in \iota^{-1}\kappa\Delta^{(k)} = R^k$ $(j \in \mathbb{Z})$, there exists a solution $y : \mathbb{R} \longrightarrow \mathbb{R}$ with $y_0 = l_0 + \psi_0 = -1 + \psi_0 \in -1 + R^k$ which follows the level sequence **l**. Assertion a) is proved, since $\mathbf{l} \in \mathcal{L}$ was arbitrary. $\square$

It comes as no surprise that the robustness statement of Corollary 3.4 leads to a corresponding result for functional differential equations 'close' to equation $(\hat{f})$ in the above situation.

4.15 Theorem. *Under the conditions of Theorem 4.14, let $k^* \in \mathbb{N}$ and the sets R^k $(k \geq k^*)$ be as in assertion a) of that theorem. There exists $k^{**} \in \mathbb{N}$, $k^{**} \geq k^*$ such that for all $k_1 \in \mathbb{N}, k_1 \geq k^{**}$ there exists $\theta(k_1) > 0$ with the following property:*

Assume that $F \in BC^1(C, \mathbb{R})$ satisfies the conditions

$$F(\psi + 2) = F(\psi) \quad (\psi \in C),$$

$$\|F - \hat{f}\|_l \leq \theta(k_1),$$

and conditions (ii, $\theta(k_1)$) and (4.8.1) of Proposition 4.8.

Then the statements analogous to a) and c) of Theorem 4.15 hold for equation (F) on each set $-1 + R^k$, $k^{**} \leq k \leq k_1$.

Proof. Recall that $R^k = \iota^{-1}\kappa\Delta^{(k)}$ and $\mathcal{T}_1 = \mathcal{T}_{1,\hat{f}}$. We know (compare the proof of Theorem 4.14) that $\mathcal{X} := \kappa^{-1}\mathcal{T}_1\mathcal{T}_0\kappa$ is C^1 on a neighborhood of $\Delta^{(k)}$ $(k \geq k^*)$. Applying Corollary 3.4, we obtain (for every $k \geq k^*$) a number $\varepsilon_k > 0$ with the following property: If $F \in BC^1(C, \mathbb{R})$ is such that

(4.15.1) $\quad \mathcal{T}_{1,F} \circ \mathcal{T}_{0,R^k,F}$ is defined on an open neighborhood W of $\kappa(\Delta^{(k)})$ $(= \iota(R^k))$ in C, and
$$\|(\mathcal{T}_{1,F} \circ \mathcal{T}_{0,R^k,F} - \mathcal{T}_1 \circ \mathcal{T}_0)_{|W}\|_{C^1} < \varepsilon_k,$$

then the map $\chi_F := \kappa^{-1} \circ \mathcal{T}_{1,F} \circ \mathcal{T}_{0,R^k,F} \circ \kappa$ is defined on a neighborhood of $\Delta^{(k)}$ and also satisfies the conditions of Theorem 2.4 on $\Delta^{(k)}$.

From (4.14.4), we have $R^k \subset \mathcal{K} \cap U_L$ for $k \geq k^*$. Recall now $z_1 > 0$ from Proposition 4.10. In view of Theorem 4.14, b), there exists $k^{**} \geq k^*$ such that
$$\forall k \geq k^{**} \, \forall \psi \in R^k : |\mathrm{pr}_3 \iota \psi| < z_1.$$

Fix now $k \geq k^{**}$. Then Proposition 4.10, applied with $Q := R^k$, shows that there exists $\theta(\varepsilon_k)$ such that if $F \in BC^1(C, \mathbb{R})$ satisfies $\|F - \hat{f}\|_l \leq \theta(\varepsilon_k)$ and conditions (ii, $\theta(\varepsilon_k)$) and (4.8.1) of Proposition 4.8, then $F \in \mathcal{B}_1^+ \cap \mathcal{B}_1^- \cap \mathcal{B}_- \cap \mathcal{B}_Q$, and (4.15.1) holds. (With $\tilde{W}_Q = \tilde{W}_{R^k}$ as in Proposition 4.10, set $W := \iota \tilde{W}_{R^k}$.) Let now F be as in the assertion of the present theorem; that is, F satisfies, in addition, $F(\psi + 2) = F(\psi)$ ($\psi \in C$). Then F satisfies condition (4.11.1), and hence Proposition 4.12 applies. The proof of assertions c) and a) of Theorem 4.14 is then valid with $\hat{f}$ replaced by F, and $\mathcal{T}_1, \mathcal{T}_0$ replaced by $\mathcal{T}_{1,F}$ and $\mathcal{T}_{0,R^k,F}$. It follows that the statements of Theorem 4.14 hold for equation (F) on R^k.

The assertion of the present theorem is now obtained by setting $\theta(k_1) := \min\{\theta(\varepsilon_k) \mid k^{**} \leq k \leq k_1\}$. $\square$

Remarks on the linearity condition (f2) for the results of this section. This condition is restrictive, of course. One possible step towards a generalization of Theorem 4.14 would be to prove an analog for delay equations of Propositions 3.2.5 and 3.2.6 (p. 197) from [Wiggins]. (These results express C^1-closeness of the 'local' map $\mathcal{T}_0$ to the corresponding map in the nonlinear case.)

We did not pursue this in the present work because of the additional technical effort that would be necessary, and because of the subsequent reasons:

1) An analog of condition (v) in a not locally linear situation would be the following: With $a := \mathrm{Re}(v)$, $b := \mathrm{Im}(v)$, the space spanned by a_{t_1} and b_{t_1} is not contained in the tangent space to the stable manifold of 1 at x_{t_1}. One would need a criterion for this property in terms of zeroes of a and b – a result in the spirit of the transversality criterion from [Lani-Wayda, Walther 1], which applies to periodic solutions under negative feedback. Here, a corresponding result for equilibria and with locally positive feedback would be required. Such a tool is currently not available.

2) Even with such a criterion, the verification in case of a concrete example would require rather explicit knowledge of the heteroclinic solution x. Presently we have no such example, except for the piecewise linear one.

3) We obtain C^1-smooth examples by perturbation from a function f satisfying (f1)–(f3).

5. Explicit examples

In this section we present the main results - an example of a delay equation to which Theorem 4.14 applies, and perturbations of this example which satisfy the conditions of Theorem 4.15. Thus we obtain the existence of erratic solutions, describable by symbolic dynamics. The results are, to the author's knowledge, currently the most explicit examples for 'chaos' in scalar delay equations – in the sense of simple analytical expressions for the nonlinearities, which are not constant on any interval. The examples are also new in the following respect: Erratic solutions do *not* arise from a periodic and a transversally homoclinic solution, as in all previous examples ([An der Heiden, Walther], [Hale, Lin], [Lani-Wayda, Walther 1], [Lani-Wayda, Walther 2], [Lani-Wayda 4]). The relevant mechanism is rather heteroclinic solutions between equilibria. The (infinite–dimensional) situation is somewhat similar to a spiral saddle in $\mathbb{R}^3$ with *both* branches of the unstable manifold homoclinic to the saddle; compare, e.g., [Wiggins], p. 253-257. In this 3–dimensional setting, a result analogous to the ones from this section would be existence of solutions close to the union of the two homoclinic branches, following the one or the other branch in an unpredictably alternating manner.

Solutions of delay equations that go up and down between different levels in a seemingly random fashion were observed in numerical experiments with the equation

$$\dot{x}(t) = -\alpha \sin(x(t-1)),$$

see [Dormayer, Lani-Wayda]. Let us consider a piecewise linear caricature of the sine function, or rather of the function $x \mapsto \frac{1}{\pi}\sin(\pi x)$. Our aim is to explain quasi–random solution behavior, at least in the sense of a proof of its *existence*. We can presently not give a rigorous explanation for the fact that, for large enough α, *most* numerically observed solutions show such behavior. Define $\check{s}: \mathbb{R} \to \mathbb{R}$ by

$$\check{s}(x) := \frac{1}{2} - |x - \frac{1}{2}| \text{ for } x \in [-1/2, 3/2],$$

and by $\check{s}(x+2) = \check{s}(x)$ ($x \in \mathbb{R}$). (See Figure 5.) We consider the one-parameter family of equations

$$(-\alpha\check{s}) \qquad \dot{x}(t) = -\alpha\check{s}(x(t-1)) \quad (\alpha > 0).$$

Note that $\check{s}$ is odd, and $-\check{s}(2k-1+x) = x$ for $|x| \leq \frac{1}{2}$.

In particular, for all equilibria $2k-1$ ($k \in \mathbb{Z}$) of equation $(-\alpha\check{s})$, the linear variational equation about this equilibrium is

$$(\alpha) \qquad \dot{v}(t) = \alpha v(t-1).$$

The following statement is obvious:

5.1. Proposition. *If $\alpha \in (0, 3\pi/2)$ then $f = (-\alpha \check{s})$ satisfies conditions (f1) and (f2) from Section 4, with $\bar{l} = 1/2$.*

It is clear that, if we want to apply Theorem 4.14, we have to look for heteroclinic solutions of $(-\alpha \check{s})$, joining the equilibria -1 and $+1$. (In the development of this work, the heteroclinic solutions were found first, and the framework for the description of erratic solutions was constructed later.) For $\alpha \in (0, 3\pi/2)$, one branch of the unstable manifold of -1 (associated with the characteristic value $\lambda > 0$ of equation (α)) consists of the orbit of the solution $x^\alpha : \mathbb{R} \to \mathbb{R}$ of equation $(-\alpha \check{s})$ with

$$x^\alpha(t) = -1 + \frac{1}{2} \cdot e^{\lambda t} \quad (t \leq 0),$$

where $\alpha = \lambda e^\lambda$. Our approach is simple: We try to find $\alpha > 0$ such that the segment x_2^α at time 2 is in the stable space of the equilibrium $+1$, i.e.,

(5.1) $$\mathrm{pr}_\lambda(x_2^\alpha - 1) = 0.$$

(Here pr_λ denotes the coordinate of the spectral projection onto the one-dimensional unstable space $\mathbb{R} \cdot \exp(\lambda \cdot)$ of equation (α) which is associated to $\lambda > 0$; compare Lemma 6.8.) If we find such an α, we have a *candidate* for a heteroclinic solution, but it still has to be shown that the segments x_t^α satisfy

$$|x_t^\alpha - 1| \leq \frac{1}{2} \text{ for all } t \geq 2,$$

i.e., remain in the linear region around $+1$. (Only then we can conclude from (5.1) that $x_t^\alpha \to 1$ $(t \to \infty)$.) The proof of this property is the most difficult part in the proof of the following lemma. Part of the difficulty lies in the calculation of numerical estimates, which we deferred to Proposition 6.9 in the appendix.

5.2. Lemma. *Set $\lambda^* := \log(9) - 1$ and $\alpha^* := \lambda^* \exp(\lambda^*) = (9/e)(\log 9 - 1)$.*

a) Then $\alpha^* \in (0, 3\pi/2)$, and the condition $|\rho_1| < \lambda^*$ from (f3) (Section 4) on the solution $\mu_1 = \rho_1 + i\omega_1$ of $z = \alpha^* \exp(-z)$ with largest negative real part is satisfied.

b) For $\alpha = \alpha^*$, the solution x^α of equation $(-\alpha \check{s})$ satisfies conditions (x1)-(x4) of Section 4, with $t_1 := 2$ and $l := 0.47$. (In particular, x^α is heteroclinic between -1 and 1.)

Proof. Part a) of the lemma follows from the numerical estimates in Proposition 6.9, a), d) and e).

For $\lambda > 0$, set $\alpha(\lambda) = \lambda \exp(\lambda)$, so that λ is the unique real solution of the characteristic equation $z = \alpha(\lambda) \exp(-z)$ associated with equation $(\alpha(\lambda))$. (Of course, $\alpha^* = \alpha(\lambda^*)$.)

Proof of b): Let us first compute the solution $x^{\alpha(\lambda)}$ of $(-\alpha \check{s})$ on $[0, 2]$ under the assumption

(5.2.1) $$\log 5 > \lambda > \log 3.$$

We write x instead of $x^{\alpha(\lambda)}$. We have $x(t) = -1 + \frac{1}{2}e^{\lambda t}$ for $t \leq 0$, and equation $(-\alpha \check{s})$ shows that even

(5.2.2) $$x(t) = -1 + \frac{1}{2}e^{\lambda t} \text{ for } t \in (-\infty, 1].$$

Condition (5.2.1) implies that with the (λ–dependent) numbers $t_0 := \frac{1}{\lambda} \log 2$, $t_{1/2} := \frac{1}{\lambda} \log 3$ we have

$$0 < t_0 < t_{1/2} < 1, \quad x(t_0) = 0, \quad x(t_{1/2}) = 1/2.$$

On $[1, t_{1/2} + 1]$,

$$\dot{x}(t) = -\alpha x(t-1) = -\alpha(-1 + \frac{1}{2}e^{\lambda(t-1)}) = \alpha - \frac{1}{2}\lambda e^{\lambda t}, \text{ and}$$

(5.2.3)
$$\begin{aligned} x(t) &= x(1) + \int_1^t (\alpha - \frac{1}{2}\lambda e^{\lambda s}) ds \\ &= -1 + \frac{1}{2}e^{\lambda} + \alpha(t-1) - [\frac{1}{2}e^{\lambda s}]_1^t \\ &= -1 + \frac{1}{2}e^{\lambda} + \alpha(t-1) - \frac{1}{2}e^{\lambda t} + \frac{1}{2}e^{\lambda} \\ &= -1 + \frac{\alpha}{\lambda} + \alpha(t-1) - \frac{1}{2}e^{\lambda t}. \end{aligned}$$

Thus
$$\begin{aligned} x(t_{1/2} + 1) &= -1 + \frac{\alpha}{\lambda} + \alpha(\frac{1}{\lambda}\log 3 - 1) - \frac{1}{2}e^{\log 3 + \lambda} \\ &= -1 + \frac{\alpha}{\lambda}(1 + \log 3) - \frac{3}{2}\frac{\alpha}{\lambda} \\ &= -1 + \frac{\alpha}{\lambda}(\log 3 - \frac{1}{2}). \end{aligned}$$

On $[t_{1/2} + 1, 2]$, we have from (5.2.1) that $x(t-1) \in [1/2, 3/2]$, and

$$\dot{x}(t) = \alpha(-1 + \frac{1}{2}e^{\lambda(t-1)} - 1) = \alpha(-2 + \frac{1}{2}e^{\lambda(t-1)}) = -2\alpha + \frac{1}{2}\lambda e^{\lambda t}, \text{ and}$$

(5.2.4)
$$\begin{aligned} x(t) &= x(t_{1/2} + 1) + [(-2\alpha s + \frac{1}{2}e^{\lambda s})]_{t_{1/2}+1}^t \\ &= -1 + \frac{\alpha}{\lambda}(\log 3 - \frac{1}{2}) + \frac{1}{2}e^{\lambda t} - \frac{1}{2} \cdot 3\frac{\alpha}{\lambda} - 2\alpha(t - t_{1/2} - 1) \\ &= -1 + \frac{\alpha}{\lambda}(\log 3 - 2) + \frac{1}{2}e^{\lambda t} - 2\alpha(t - t_{1/2} - 1) \\ &= -1 + \frac{\alpha}{\lambda}(3\log 3 - 2) + 2\alpha - 2\alpha t + \frac{1}{2}e^{\lambda t}. \end{aligned}$$

EXPLICIT EXAMPLES

We have

(5.2.5) $\qquad \dot{x} > 0$ on $[1, t_0+1)$, $\quad \dot{x} < 0$ on $(t_0+1, 2]$.

Setting $t_{\max} = t_0 + 1$, we compute

(5.2.6)
$$x(t_{\max}) = -1 + \frac{\alpha}{\lambda} + \alpha \cdot t_0 - \frac{1}{2}e^{\lambda(t_0+1)} = -1 + \frac{\alpha}{\lambda} + \frac{\alpha}{\lambda}\log 2 - \frac{\alpha}{\lambda}$$
$$= -1 + \frac{\alpha}{\lambda}\log 2.$$

Further,

(5.2.7)
$$x(1) = -1 + \frac{1}{2}e^\lambda = -1 + \frac{1}{2}\frac{\alpha}{\lambda},$$
$$x(2) = -1 + \frac{\alpha}{\lambda}(\log 3 - 2) + \frac{1}{2}(\frac{\alpha}{\lambda})^2 - 2\alpha(1 - \frac{1}{\lambda}\log 3)$$
$$= -1 + \frac{\alpha}{\lambda}(3\log 3 - 2) - 2\alpha + \frac{1}{2}(\frac{\alpha}{\lambda})^2.$$

We now define $y^\lambda : [-1, \infty) \to \mathbb{R}$ by
$$y_0^\lambda = x_2^{\alpha(\lambda)} - 1, \text{ and } \dot{y}^\lambda(t) = \alpha y^\lambda(t-1) \quad (t \geq 0).$$

Claim: The number $\text{pr}_\lambda y_0^\lambda$ describing the projection of y_0^λ to the unstable space $\mathbb{R} \cdot \exp(\lambda \cdot)$ of equation (α) satisfies

(5.2.8) $\qquad \text{pr}_\lambda y_0^\lambda = \frac{1}{2}\frac{\exp(2\lambda)(\lambda - \lambda^*)}{1 + \lambda}.$

Proof: In the calculation, we write y for y^λ and α for $\alpha(\lambda)$; note that then $e^\lambda = \alpha/\lambda$. Using the formula from Lemma 6.8, b) we get

$$\text{pr}_\lambda y_0^\lambda = \frac{1}{1+\lambda}[y(0) + \lambda \underbrace{\int_{-1}^0 e^{-\lambda s} y(s) ds}_{=:I}],$$

$$I = [-e^{-\lambda s} y(s)]_{-1}^0 + \underbrace{\int_{-1}^0 e^{-\lambda s} \dot{y}(s) ds}_{=:J}, \text{ and}$$

$$J = \int_{-1}^0 e^{-\lambda s} \dot{x}(2+s) ds = -\alpha \int_{-1}^0 e^{-\lambda s} \check{s}(x(1+s)) ds$$
$$= -\alpha \int_0^1 \frac{\alpha}{\lambda} e^{-\lambda s} \check{s}(x(s)) ds$$
$$= -\frac{\alpha^2}{\lambda}\{\int_0^{t_{1/2}} e^{-\lambda s}(-1 + \frac{1}{2}e^{\lambda s}) ds + \int_{t_{1/2}}^1 e^{-\lambda s}(2 - \frac{1}{2}e^{\lambda s}) ds\}$$
$$= -\frac{\alpha^2}{\lambda}\{-\frac{1}{\lambda}e^{-\lambda s}|s = t_{1/2}(-1-2) - \frac{1}{\lambda} - 2\frac{e^{-\lambda}}{\lambda} + \frac{1}{2}(t_{1/2} - (1 - t_{1/2}))\}$$
$$= -\frac{\alpha^2}{\lambda}\{\frac{1}{\lambda} - \frac{1}{\lambda} - \frac{2}{\alpha} + \frac{1}{\lambda}\log 3 - \frac{1}{2}\} = -\frac{\alpha^2}{\lambda} \cdot \frac{2\log 3 - \lambda}{2\lambda} + \frac{2\alpha}{\lambda}.$$

Thus, $y(0) + I = e^\lambda y(-1) + J = e^\lambda(-2 + \frac{\alpha}{2\lambda}) + J$ (see (5.2.7)), and

$$\mathrm{pr}_\lambda y_0^\lambda = \frac{1}{1+\lambda} \cdot \frac{\alpha^2}{\lambda^2}(\frac{1}{2} - \frac{2\log 3 - \lambda}{2}) = \frac{\exp(2\lambda)(\lambda - (\log 9 - 1))}{2(1+\lambda)}.$$

Since $\lambda^* = \log 9 - 1$, formula (5.2.8) is proved. We have $\lambda^* = \log 3 + \log 3 - 1 > \log 3$, and $\lambda^* < \log 5$, since $e^{\lambda^*} = 9/e < 5$. Hence condition (5.2.1) holds for $\lambda = \lambda^*$, and we conclude that

(5.2.9) $$\mathrm{pr}_{\lambda^*} y_0^{\lambda^*} = \mathrm{pr}_{\lambda^*}(x_2^{\alpha^*} - 1) = 0.$$

We set $\lambda := \lambda^*$ and $\alpha := \alpha^*$ from now on, $t_0 := \frac{1}{\lambda^*}\log 2$, $t_{1/2} := \frac{1}{\lambda^*}\log 3$, and we write y for y^{λ^*}.

(5.2.10) *Claim:* $|y(t)| \leq 0.46$ for all $t \in [-1, \infty)$.

Proof: With $q := \dfrac{\alpha}{\lambda} = e^\lambda = \dfrac{9}{e}$, we obtain from (5.2.3) and (5.2.4) that

(5.2.11)
$$y(t) = -2 + q + \alpha(t+1) - \frac{1}{2}q^2 e^{\lambda t}$$
$$= -2 + e^\lambda + \lambda e^\lambda(t+1) - \frac{1}{2}e^{\lambda(2+t)}$$

for $t \in [-1, t_{1/2} - 1]$, and for $t \in [t_{1/2} - 1, 0]$ we have

(5.2.12)
$$y(t) = -2 + q(3\log 3 - 2) - 2\alpha(t+1) + \frac{1}{2}q^2 e^{\lambda t},$$
$$= -1 + e^\lambda(3\log 3 - 2) - 2\lambda e^\lambda(t+1) + \frac{1}{2}e^{\lambda(2+t)}.$$

Further, from (5.2.5)-(5.2.7), we see that with $\tilde{t}_{\max} := -2 + t_{\max} = t_0 - 1 = \frac{1}{\lambda}\log 2 - 1$, one has $\dot y > 0$ on $[-1, \tilde{t}_{\max})$, $\dot y < 0$ on $(\tilde{t}_{\max}, 0]$, and

(5.2.13)
$$\begin{aligned} y(-1) &= -2 + q/2, \\ y(0) &= -2 + q(3\log 3 - 2) - 2\alpha + q^2/2 \\ &= -2 + q(3\log 3 - 2 - 2\lambda) + q^2/2 \\ &= -2 - q\log 3 + q^2/2, \\ y(\tilde{t}_{\max}) &= -2 + q\log 2. \end{aligned}$$

Using the numerical estimates from Proposition 6.9, a)–c), we infer

$$0 \geq y(-1) \geq -2 + \frac{3.31}{2} \geq -0.35,$$
$$y(\tilde{t}_{\max}) \leq -2 + 3.32 \cdot 0.7 \leq -2 + 2.324 = 0.324,$$
$$y(0) \geq -2 - 3.32 \cdot 1.1 + 3.3^2/2$$
$$= -2 - 3.652 + 10.89/2$$
$$= -5.652 + 5.445 = -0.207.$$

Hence we have

(5.2.14) $$\max_{t\in[-1,0]} |y(t)| \leq 0.35.$$

It remains to prove (5.2.10) for $t > 0$. Recall that $\mathrm{pr}_\lambda y_0 = 0$. With μ_k ($k \in \mathbb{N}$) as in Lemma 6.8, we have for $t > 0$ the series expansion (see Lemma 6.8, d))

$$y(t) = \sum_{k=1}^{\infty} \mathrm{pr}_{\mu_k} y_0 \cdot \exp(\mu_k t) + \mathrm{pr}_{\bar{\mu}_k} y_0 \cdot \exp(\bar{\mu}_k t).$$

Since y_0 is real, we have $\mathrm{pr}_{\bar{\mu}_k} y_0 = \overline{\mathrm{pr}_{\mu_k} y_0}$ for $k \in \mathbb{N}$, and hence, since $\mathrm{Re}(\mu_k) < 0$,

(5.2.15) $$|y(t)| \leq 2 \sum_{k=1}^{\infty} |\mathrm{pr}_{\mu_k} y_0| \quad (t > 0).$$

For $\mu \in \{\mu_1, \mu_2 ...\}$, we now compute the number $\mathrm{pr}_\mu y_0$ (recall $e^\mu = \frac{\alpha}{\mu}$).

$$\mathrm{pr}_\mu y_0 = \frac{1}{1+\mu}\{y(0) + \underbrace{\mu \int_{-1}^{0} e^{-\mu s} y(s) ds}_{=:I}\},$$

$$I = [-e^{-\mu s} y(s)]_{-1}^{0} + \underbrace{\int_{-1}^{0} e^{-\mu s} \dot{y}(s) ds}_{=:J}, \text{ and}$$

$$J =$$
$$= -\frac{\alpha^2}{\mu}\{\int_{0}^{t_{1/2}} e^{-\mu s}(-1 + \frac{1}{2}e^{\lambda s})ds + \int_{t_{1/2}}^{1} e^{-\mu s}(2 - \frac{1}{2}e^{\lambda s})ds\}$$
$$= -\frac{\alpha^2}{\mu}\{-\frac{1}{\mu}e^{-\mu t_{1/2}}(-3) - \frac{1}{\mu} - \frac{2}{\alpha} + [\frac{1}{2(\lambda-\mu)}e^{(\lambda-\mu)s}]_0^{t_{1/2}} -$$
$$[\frac{1}{2(\lambda-\mu)}e^{(\lambda-\mu)s}]_{t_{1/2}}^{1}\}$$
$$= -\frac{\alpha^2}{\mu}\{\frac{3}{\mu}e^{-\frac{\mu}{\lambda}\log 3} - \frac{1}{\mu} - \frac{2}{\alpha} + \frac{1}{2(\lambda-\mu)}[2e^{\frac{\lambda-\mu}{\lambda}\log 3} - 1 - \frac{\mu}{\lambda}]\}.$$

$$y(0) + I =$$
$$= e^{\mu}(-2 + \frac{\alpha}{2\lambda}) + J$$
$$= \frac{\alpha}{\mu}\{-2 + \frac{\alpha}{2\lambda} - \frac{3\alpha}{\mu}e^{-\frac{\mu}{\lambda}\log 3} + \frac{\alpha}{\mu} + 2 - \frac{\alpha \cdot 3e^{-\frac{\mu}{\lambda}\log 3}}{\lambda - \mu} + \frac{\alpha(\lambda + \mu)}{2(\lambda - \mu)\lambda}\}$$
$$= \frac{\alpha}{\mu} \cdot$$
$$\frac{\alpha\mu(\lambda - \mu) - 6\alpha\lambda(\lambda - \mu)e^{-\frac{\mu}{\lambda}\log 3} + 2\alpha\lambda(\lambda - \mu) - 6\alpha\lambda\mu e^{-\frac{\mu}{\lambda}\log 3} + \mu\alpha(\lambda + \mu)}{2\lambda\mu(\lambda - \mu)}$$
$$= \frac{\alpha^2}{2\mu^2\lambda(\lambda - \mu)} \cdot (-6\lambda^2 e^{-\frac{\mu}{\lambda}\log 3} + 2\lambda^2)$$
$$= \frac{\alpha^2 \lambda(1 - 3e^{-\frac{\mu}{\lambda}\log 3})}{\mu^2(\lambda - \mu)} = \frac{\alpha\lambda(e^{\mu} - 3e^{\mu(1-\frac{\log 3}{\lambda})})}{\mu(\lambda - \mu)}.$$

Hence we obtain

(5.2.16) $$\mathrm{pr}_{\mu} y_0 = \frac{\alpha\lambda(e^{\mu} - 3e^{\mu(1-\frac{\log 3}{\lambda})})}{\mu(1 + \mu)(\lambda - \mu)}.$$

From Lemma 6.8, a), we have $\mu_k = \rho_k + i\omega_k$, with $\rho_k < 0$ and $\omega_k \in (2k-1)\pi, 2k\pi)$ ($k \in \mathbb{N}$). Using the estimate from Proposition 6.9, d) on $\alpha\lambda$, we obtain for $k \in \mathbb{N}$ from (5.2.16) that

(5.2.17) $$2|\mathrm{pr}_{\mu_k} y_0| \leq \frac{2 \cdot 4.77 \cdot 4}{[\mathrm{Im}(\mu_k)]^3} \leq \frac{2 \cdot 19.08}{(2k-1)^3 \pi^3} \leq \frac{2 \cdot 0.616}{(2k-1)^3} = \frac{1.232}{(2k-1)^3},$$

where we have used the (hand–calculated) estimates $\pi^3 \geq 3.14159^3 \geq 31.002$ and $19.08/31.002 \leq 0.616$.

We first estimate the terms for $k \geq 2$ in (5.2.15):

$$\sum_{k=2}^{\infty} \frac{1}{(2k-1)^3} = \frac{1}{27} + \frac{1}{125} + \frac{1}{343} + \sum_{k=5}^{\infty} \frac{1}{(2k-1)^3} \leq$$
$$\leq \frac{55511}{1,157,625} + \int_4^{\infty} \frac{1}{(2x-1)^3} dx =$$
$$= \ldots + [\frac{1}{4(2x-1)^2}]|x=4$$
$$= \ldots + \frac{1}{196} \leq 0.048 + 0.0052$$
$$= 0.0532$$

(where the last estimates are calculated by hand). It follows that

(5.2.18) $$2\sum_{k=2}^{\infty} |\mathrm{pr}_{\mu_k} y_0| \leq 1.232 \cdot 0.0532 \leq 0.066.$$

Now we estimate the term for $k = 1$: From Proposition 6.9, e), we know that $\omega_1 \geq 4.67$. In view of the first inequality in (5.2.17), we get

$$(5.2.19) \qquad 2|\mathrm{pr}_{\mu_1} y_0| \leq 2 \cdot \frac{19.08}{\omega_1^3} \leq \frac{38.16}{4.67^3} \leq \frac{38.16}{100} \leq 0.39.$$

Combining (5.2.19), (5.2.18) and (5.2.15), we obtain for $t > 0$

$$(5.2.20) \qquad |y(t)| \leq 0.39 + 0.066 \leq 0.46.$$

Claim (5.2.10) follows from the last estimate and (5.2.14).

We can now finish the proof of part b). Define $\bar{x}(t) := x(t)$ for $t \leq 2$, and $\bar{x}(t) := 1 + y(t - 2)$ for $t > 2$. The definition implies $\bar{x}(t - 1) = 1 + y(t - 3)$ for $t \geq 3$, and for $t \in [2, 3]$ this identity holds since $\bar{x}_2 = x_2 = 1 + y_0$. The function $\bar{x} : \mathbb{R} \to \mathbb{R}$ is a solution of equation $(-\alpha \check{s})$, since x is a solution, and since for $t > 2$ one has (from the definition of $\check{s}$ and from (5.2.10))

$$\dot{\bar{x}}(t) = \dot{y}(t-2) = \alpha y(t-3) = -\alpha \check{s}(1 + y(t-3)) = -\alpha \check{s}(\bar{x}(t-1)).$$

It follows that $x = \bar{x}$. We check conditions (x1)-(x4) with $\bar{l} := 1/2$, $l := 0.47$ and $t_1 := 2$:

Properties (x1) and (x2) follow from (5.2.2). Property (x3) follows from $x = \bar{x}$, together with $x_2 = 1 + y_0$ and (5.2.10).

Proof of property (x4): If we had $\mathrm{pr}_2(x_{t_1} - 1) = 0$, which in our situation is equivalent to $\mathrm{pr}_{\mu_1} y_0 = 0$, then (5.2.16) would imply

$$1 - 3e^{-\mu_1 \frac{\log 3}{\lambda}} = 0, \text{ or } e^{(1 - \frac{\mu_1}{\lambda}) \log 3} = 1,$$

so that $\mathrm{Re}(\mu_1) = \lambda$; but we know $\mathrm{Re}(\mu_1) < 0 < \lambda$. Hence $\mathrm{pr}_2(x_{t_1} - 1) \neq 0$, and the lemma is proved. □

5.3. Lemma. *The complex valued solution of the variational equation $(-\alpha^* \check{s}, x^{\alpha^*})$ with initial value $\exp(\mu_1 \cdot)$ satisfies condition (v) of Section 4.*

Proof. With $\alpha = \alpha^*$, we have to compute the solution of the initial value problem
$$v(t) = \exp(\mu_1 t) \quad (t \in [-1, 0])$$
$$\dot{v}(t) = -\alpha \check{s}'(x^\alpha(t-1))v(t-1)$$

on the interval $[0, 2]$, and to show that, with $\lambda = \lambda^*$,

$$(5.3.1) \qquad v(2) + \lambda \int_{-1}^{0} e^{-\lambda s} v(2+s) ds \neq 0.$$

The coefficient $a(t) := -\alpha \check{s}'(x^\alpha(t))$ is defined for all $t \in \mathbb{R}$ except for $t = 0$ and $t = t_{1/2} = \frac{1}{\lambda} \log 3$, and the linear variational equation is to be understood

in the sense of Definition and Remark 6.1 from the appendix. One sees from (5.2.2) and the definition of š that

$$a(t-1) = \begin{cases} \alpha & \text{for } t \in [0,1) \\ -\alpha & \text{for } t \in (1, t_{1/2}+1) \\ \alpha & \text{for } t \in (t_{1/2}+1, 2] \end{cases}$$

Since μ_1 is a solution of $z = \alpha e^{-z}$, we have

$$v(t) = \exp(\mu_1 t) \text{ also for } t \in [0,1].$$

For the further calculation, we set $\mu := \mu_1$.

For $t \in [1, t_{1/2}+1]$, we have

$$v(t) = e^\mu + \int_1^t -\alpha e^{\mu(s-1)}\,ds = e^\mu - \int_1^t \mu e^{\mu s}\,ds = 2e^\mu - e^{\mu t}.$$

For $t \in [t_{1/2}+1, 2]$,

$$\begin{aligned} v(t) &= 2e^\mu - e^{\frac{\mu}{\lambda}\log 3 + \mu} + \alpha \int_{t_{1/2}+1}^t e^{\mu(s-1)}\,ds \\ &= e^\mu[2 - e^{\frac{\mu}{\lambda}\log 3}] + e^{\mu t} - e^{\frac{\mu}{\lambda}\log 3 + \mu} \\ &= 2e^\mu[1 - e^{\frac{\mu}{\lambda}\log 3}] + e^{\mu t}. \end{aligned}$$

We start computing the expression in (5.3.1):

$$\begin{aligned} \lambda \int_{-1}^0 e^{-\lambda s} v(2+s)\,ds &= \lambda \int_1^2 e^{-\lambda(s-2)} v(s)\,ds \\ &= e^{2\lambda} \left\{ \lambda \int_1^{t_{1/2}+1} e^{-\lambda s} v(s)\,ds + \lambda \int_{t_{1/2}+1}^2 e^{-\lambda s} v(s)\,ds \right\} \\ &= e^{2\lambda}\{I_1 + I_2\}, \end{aligned}$$

where I_1 is the first and I_2 the second summand in the bracket. Now

$$\begin{aligned} I_1 &= 2e^\mu[-e^{-\lambda s}]_1^{1+\frac{\log 3}{\lambda}} - \frac{\lambda}{\mu-\lambda}[e^{(\mu-\lambda)s}]_1^{1+\frac{\log 3}{\lambda}} \\ &= 2e^{\mu-\lambda} \cdot [1 - \frac{1}{3}] - \frac{\lambda}{\mu-\lambda} e^{\mu-\lambda} \cdot [\frac{e^{\frac{\mu}{\lambda}\log 3}}{3} - 1] \\ &= e^{\mu-\lambda}[\frac{4}{3} - \frac{\lambda}{\mu-\lambda}(\frac{e^{\frac{\mu}{\lambda}\log 3}}{3} - 1)]. \end{aligned}$$

Recalling that $e^\lambda = \frac{9}{e}$, we compute that, with $K := 2e^\mu(1 - e^{\frac{\mu}{\lambda}\log 3})$,

$$I_2 = K \cdot \int_{t_{1/2}+1}^{2} \lambda e^{-\lambda s} ds + \int_{t_{1/2}+1}^{2} \lambda e^{(\mu-\lambda)s} ds$$

$$= K \cdot [-e^{-\lambda s}]_{1+\frac{\log 3}{\lambda}}^{2} + \frac{\lambda}{\mu-\lambda}[e^{(\mu-\lambda)s}]_{1+\frac{\log 3}{\lambda}}^{2}$$

$$= K \cdot \frac{e}{9} \cdot [\frac{1}{3} - \frac{e}{9}] + \frac{\lambda}{\mu-\lambda}e^{\mu-\lambda} \cdot [e^{\mu-\lambda} - \frac{e^{\frac{\mu}{\lambda}\log 3}}{3}].$$

Thus, $I_1 + I_2 = K \cdot \frac{e}{9}[\frac{1}{3} - \frac{e}{9}] + e^{\mu-\lambda}\{\frac{4}{3} + \frac{\lambda}{\mu-\lambda}[e^{\mu-\lambda} - \frac{2 \cdot e^{\frac{\mu}{\lambda}\log 3}}{3} + 1]\}$, and

$$e^{2\lambda}(I_1 + I_2)$$
$$= \frac{9^2}{e^2}(I_1 + I_2)$$
$$= K \cdot \frac{9}{e}\frac{1}{3}[\frac{1}{3} - \frac{e}{9}] + e^\mu \cdot \{\frac{12}{e} + \frac{\lambda}{\mu-\lambda}[e^\mu - \frac{6e^{\frac{\mu}{\lambda}\log 3}}{e} + \frac{9}{e}]\}$$
$$= e^\mu \cdot \{2(1 - e^{\frac{\mu}{\lambda}\log 3})(\frac{3}{e} - 1) + \frac{12}{e} + \frac{\lambda}{\mu-\lambda}[e^\mu - \frac{6}{e}e^{\frac{\mu}{\lambda}\log 3} + \frac{9}{e}]\}.$$

Finally, the expression in (5.3.1) equals

(5.3.2)
$$e^\mu\{\frac{6}{e}(1 - e^{\frac{\mu}{\lambda}\log 3}) + e^\mu + \frac{12}{e} + \frac{\lambda}{\mu-\lambda}[e^\mu - \frac{6}{e}e^{\frac{\mu}{\lambda}\log 3} + \frac{9}{e}]\}$$
$$= e^\mu\{\frac{18}{e} + \frac{\mu}{\mu-\lambda}[e^\mu - \frac{6}{e}e^{\frac{\mu}{\lambda}\log 3}] + \frac{9\lambda}{e(\mu-\lambda)}\}.$$

From Proposition 6.9, a), d), we see that $\lambda \leq 2$ and $\omega_1 = \text{Im}(\mu_1) = \text{Im}(\mu) > 4$. It follows that

$$|\frac{9}{e}\frac{\lambda}{\mu-\lambda}| \leq \frac{9}{e}\frac{\lambda}{\omega_1} \leq \frac{9}{2e}.$$

Further, $\text{Re}(\mu) < 0$ implies

$$|\frac{\mu}{\mu-\lambda}[e^\mu - \frac{6}{e}e^{\frac{\mu}{\lambda}\log 3}]| \leq 1 \cdot [1 + \frac{6}{e}].$$

Using the two last estimates, we obtain for the bracket in (5.3.2)
$$|\{...\}| \geq \frac{18}{e} - \frac{e + 6 + 9/2}{e} \geq \frac{4}{e} > 0. \quad \square$$

We can now easily prove the main result. Recall the sets $\mathcal{L}$ and $\mathcal{L}^+$ of symbol sequences $\mathbf{l} = (...l_{-1}l_0l_1...) \in \mathbb{Z}^\mathbb{Z}$, respectively $\mathbf{l} = (l_0, l_1, l_2, ...) \in \mathbb{Z}^{\mathbb{N}_0}$ with $l_0 = -1, |l_{k+1} - l_k| = 2$ ($k \in \mathbb{Z}$, respectively $k \in \mathbb{N}_0$), and recall the notion of symbolic dynamics from Definition 4.2.

5.4. Theorem.

a) For $\alpha = \alpha^* = (9/e)(\log 9 - 1)$, the conditions of Theorem 4.14 are satisfied by $f = (-\alpha \check{s})$ and $x = x^\alpha$.

b) There exists $k^* \in \mathbb{N}$, and a countable collection of closed sets $R^k \subset C$ ($k \geq k^*$) as described in Theorem 4.14, b), such that equation $(-\alpha \check{s})$ has symbolic dynamics with respect to $\mathcal{L}$ on each set $-1 + R^k$ ($k \geq k^*$).

c) For every $k \geq k^*$ and every level sequence $\mathbf{l} \in \mathcal{L}^+$ there exists a function $\varphi_{k,\mathbf{l}}$ with graph $\varphi_{k,\mathbf{l}} \subset R^k$ as in Theorem 4.14, c), such that for all $\psi \in$ graph $\varphi_{k,\mathbf{l}}$ the solution $y^{-1+\psi}$ of equation $(-\alpha \check{s})$ with initial value $-1 + \psi$ follows the level sequence $\mathbf{l}$.

Proof. We know from Proposition 5.1 and from Lemma 5.2, a) that conditions (f1)-(f3) hold for $f = (-\alpha^* \check{s})$. Lemma 5.2, b) and Lemma 5.3 show that the heteroclinic solution x^{α^*} satisfies conditions (x1)-(x4), and that condition (v) holds. The assertions are obtained by application of Theorem 4.14. □

5.5. Remarks on Theorem 5.4 and related numerical observations.

Note that the sets $-1 + R^k$ ($k \geq k^*$) accumulate at -1, and that in each such set, there exist uncountably many disjoint graphs of functions $\varphi_{k,\mathbf{l}}$ ($\mathbf{l} \in \mathcal{L}^+$) which correspond to different level sequences. Obviously this is an expression of sensitive dependence on initial conditions, and the difference between solutions starting in different graphs shows itself on a 'macroscopic' scale, since any two different levels have at least distance 2.

Replacing the function $x \mapsto \frac{1}{\pi} \sin(\pi x)$ by a piecewise linear approximation allowed us to perform the explicit calculations on which the proof of Theorem 5.4 is based. One may now ask whether the phenomenon described in this theorem also occurs for the equation

$$\dot{x}(t) = -\frac{\alpha}{\pi} \sin(\pi x(t-1)), \text{ or equivalently,}$$

for the equation $\dot{y}(t) = -\alpha \sin(y(t-1))$ (replace πx by y). Consider the linear homotopy from $\check{s}$ to $\frac{1}{\pi}\sin(\pi, \cdot)$:

$$g(t, x) := (1-t)\check{s}(x) + t \cdot \frac{1}{\pi} \sin(\pi x), \quad (x \in \mathbb{R}, t \in [0,1]).$$

Starting with the α-value from Theorem 5.4 and with $t = 0$, one can numerically try to increase t from 0 to 1 in small steps and, at the same time, vary α in a way such that the heteroclinic solution is preserved. (Practically one approximates the unstable manifold of -1 by the unstable eigendirection, computes the forward solution, and determines $\alpha(t)$ for each t such that the solution oscillates about $+1$ as long as possible.) The author did this

and obtained a solution that is apparently heteroclinic between -1 and an unstable *periodic* solution oscillating about $+1$.

The explanation is simple: Since the positive 'hump' of the nonlinearity $x \mapsto -\frac{1}{\pi}\sin \pi x$ between -1 and 0 is less in size than the corresponding hump of $\check{s}$, the parameter α has to be larger for equation $(\frac{-\alpha}{\pi}\sin(\pi \cdot))$ in order to steer the solution starting in the unstable manifold of -1 up to $+1$. In fact, it turns out that α has to be larger than $3\pi/2$ to achieve this, and thus during the homotopy one enters a situation where the unstable manifold at the equilibria ± 1 is no longer one–dimensional, but three–dimensional. As the first conjugate pair of eigenvalues crosses the imaginary axis at $\alpha = 3\pi/2$, a Hopf bifurcation produces periodic solutions, which inherit the unstable eigenvalue from the equilibria. The numerically computed solution now no longer corresponds to *the* unstable manifold, but only to the one–dimensional submanifold associated with the most unstable, positive eigenvalue at -1. This solution, for appropriate α, hits the codimension-one stable manifold of the bifurcated periodic solution at $+1$. It is very likely in this situation that, for suitable values of α, one has transversally homoclinic connections between the periodic solutions of equation $(-\frac{\alpha}{\pi}\sin)$ at -1 and at $+1$. One would then have exactly the situation that is described in [Walther 3] for carefully constructed sine–like nonlinearities; however, there is no proof for this conjecture.

If one aims to reproduce the situation from Theorem 5.4 with analytic nonlinearities, one may next try a function that is closer in shape to the piecewise linear function $\check{s}$ than $\frac{1}{\pi}\sin(\pi \cdot)$. For example, the function

$$f(x) := \frac{1}{\pi}[\sin(\pi x) + (\frac{\pi}{2} - 1)(\sin(\pi x))^3]$$

has the same derivatives as $\check{s}$ at the points ± 1, and the same value $\pm \frac{1}{2}$ at the extrema $x = \pm \frac{1}{2}$ (see Figure 5). Repeating the above homotopy procedure with this f instead of $(\frac{1}{\pi}\sin(\pi \cdot))$, one finds numerically that a heteroclinic solution of the same type as for equation $(-\alpha \check{s})$ exists for an α–value less than $\frac{9}{e}(\log 9 - 1)$, namely $\alpha \approx 3.526$ (see Figure 7). Hence it seems likely that, with this α, equation $(-\alpha f)$ satisfies the conditions of Theorem 4.14 (except for the local linearity). Again, there is no proof for this conjecture, and the author has not numerically tested the analog of the transversality type condition (v) from Section 4.

In view of the above remarks it seems desirable to have at least some examples of differentiable nonlinearities to which results of Section 4 apply. We now define such functions; these examples are of class C^1 and analytic on $(0,1)$, but not on all of $\mathbb{R}$ (see also Figure 5). It is clear that, for example, sufficiently high Fourier approximations of $\check{s}$ would provide analytic nonlinearities to which Theorem 4.15 applies. We prefer to give the simpler functions below.

5.6. Definition and Remark. *For $\gamma \geq 0$, define $f_\gamma : \mathbb{R} \to \mathbb{R}$ by*

$$f_\gamma(x) := \sqrt{\frac{1}{4} + \gamma} - \sqrt{(x - \frac{1}{2})^2 + \gamma} \text{ if } x \in [0, 1],$$
$$f_\gamma(x) := -f_\gamma(-x) \text{ if } x \in [-1, 0), \text{ and}$$
$$f_\gamma(x + 2) = f_\gamma(x) \quad (x \in \mathbb{R}).$$

Then $f_0 = \check{s}$ and $f_\gamma \in C^1(\mathbb{R}, \mathbb{R})$ for $\gamma > 0$. Further,
  a) $\|f_\gamma - \check{s}\|_{C^0} \to 0$ *as $\gamma \to 0$.*
  b) *For every $\delta \in (0, 1/2)$, $f'_\gamma \to \check{s}'$ uniformly on $\mathbb{R} \setminus (\mathbb{Z} + (1/2 - \delta, 1/2 + \delta))$.*
  c) $\forall \gamma > 0 \; \forall x \in \mathbb{R} : |f'_\gamma(x)| \leq 1$.

Proof. For $\gamma = 0$, $f_\gamma(x) = \frac{1}{2} - |x - \frac{1}{2}|$ for $x \in [0, 1]$, which together with oddness and periodicity of f_0 implies $f_0 = \check{s}$. For $\gamma \geq 0$, $f_\gamma(0) = 0 = f_\gamma(1)$ implies continuity of f_γ. For $\gamma > 0$, oddness implies $\lim_{x \to 0+} f'_\gamma(x) = \lim_{x \to 0-} f'_\gamma(x)$, and oddness and periodicity show

$$\lim_{x \to 1-} f'_\gamma(x) = \lim_{x \to -1+} f'_\gamma(x) = \lim_{x \to 1+} f'_\gamma(x).$$

It follows that $f_\gamma \in C^1(\mathbb{R}, \mathbb{R})$; periodicity implies $f_\gamma \in BC^1(\mathbb{R}, \mathbb{R})$.

Proof of a): For $x \in [0, 1]$,

$$|f_\gamma(x) - \check{s}(x)| \leq |\sqrt{\frac{1}{4} + \gamma} - \frac{1}{2}| + |\sqrt{(x - \frac{1}{2})^2 + \gamma} - \sqrt{(x - \frac{1}{2})^2}|$$
$$= |\sqrt{\frac{1}{4} + \gamma} - \frac{1}{2}| + \frac{\gamma}{\sqrt{(x - \frac{1}{2})^2 + \gamma} + \sqrt{(x - \frac{1}{2})^2}}$$
$$\leq |\sqrt{\frac{1}{4} + \gamma} - \frac{1}{2}| + \frac{\gamma}{\sqrt{\gamma}} \to 0 \quad (\gamma \to 0).$$

Assertion a) follows, in view of oddness and periodicity.

Proof of b): Let $\delta \in (0, 1/2)$. For $x \in [0, 1/2 - \delta]$,

$$f'_\gamma(x) - 1 = -\frac{x - \frac{1}{2}}{\sqrt{(x - \frac{1}{2})^2 + \gamma}} - 1 = \sqrt{\frac{(x - \frac{1}{2})^2}{(x - \frac{1}{2})^2 + \gamma}} - 1$$
$$= \sqrt{1 - \frac{\gamma}{(x - \frac{1}{2})^2 + \gamma}} - 1,$$

so $0 \geq f'_\gamma(x) - 1 \geq \sqrt{1 - \frac{\gamma}{\delta^2 + \gamma}} - 1$, which shows that $f'_\gamma \to \check{s}' = 1$ $(\gamma \to 0)$ uniformly on $[0, 1/2 - \delta]$. Similarly, $f'_\gamma \to \check{s}' = -1$ uniformly on $[1/2 + \delta, 1]$, as $\gamma \to 0$. Assertion b) now also follows from the symmetries of f_γ.

Proof of c): For $x \in [0,1]$, $|f'_\gamma(x)| = \frac{|x-\frac{1}{2}|}{\sqrt{(x-\frac{1}{2})^2+\gamma}} \leq 1$. The symmetries show that this estimate holds for all $x \in \mathbb{R}$. □

Recall the number $k^{**} \in \mathbb{N}$, $k^{**} \geq k^*$ from Theorem 4.15. From the above remark and Theorem 4.15, the following last result of this section is almost obvious.

5.7. Corollary. *Set $\alpha := (9/e)(\log 9 - 1)$. For every $k_1 \in \mathbb{N}$, $k_1 \geq k^{**}$, there exists $\gamma(k_1) > 0$ such that for all $\gamma \in [0, \gamma(k_1)]$, the following holds: The equation $\dot{x}(t) = -\alpha f_\gamma(x(t-1))$ has symbolic dynamics with respect to the symbol sequences in $\mathcal{L}$ on each set R^k, $k^{**} \leq k \leq k_1$.*

The conclusions of Theorem 5.4, b), c) about solutions following symbol sequences, with initial values on graphs of Lipschitz functions, hold for equation (f_γ) on each set R^k, $k^{**} \leq k \leq k_1$.

Proof. For $\gamma > 0$, define $F_\gamma := -\alpha \hat{f}_\gamma$, i.e., $F_\gamma(\psi) = -\alpha f_\gamma(\psi(-1))$ for $\psi \in C$. Set $f := -\alpha \check{s}$, and define $\hat{f}: C \to \mathbb{R}$ correspondingly. Then $F_\gamma \in BC^1(C, \mathbb{R})$, and $F_\gamma(\psi+2) = F_\gamma(\psi)$ ($\psi \in C$). With $\theta(k_1)$ as in Theorem 4.15, and $l := 0.47$, the assertions follow from that theorem if we can find $\gamma_1 > 0$ such that for all $\gamma \in (0, \gamma_1]$,

(5.7.1) $$\|F_\gamma - \hat{f}\|_l \leq \theta(k_1), \text{ and}$$

(5.7.2) conditions (ii,$\theta(k_1)$) and (4.8.1) of Proposition 4.8 hold.

We have $\|(F_\gamma - \hat{f})|_{-1+C(l)}\|_{C^1} \leq \alpha \|(f_\gamma - \check{s})|_{(-1-l, -1+l)}\|_{C^1}$, so assertions a) and b) of 5.6, together with $l < 1/2$, show

$$\|(F_\gamma - \hat{f})|_{-1+C(l)}\|_{C^1} \to 0 \text{ as } \gamma \to 0.$$

Periodicity implies that the same property holds on $1 + C(l)$. Assertion a) of 5.6 implies $\|F_\gamma - \hat{f}\|_{C^0} \to 0$ ($\gamma \to 0$). Together, $\|F_\gamma - \hat{f}\|_l \to 0$ as $\gamma \to 0$, and there exists $\gamma_2 > 0$ such that (5.7.1) holds for all $\gamma \in (0, \gamma_2]$. Note that $\|D\hat{f}\|_\infty = \|f'\|_\infty = \sup_{x \in \mathbb{R} \setminus (\{-1/2\}+\mathbb{Z})} |f'(x)| = \alpha$. We apply Remark 4.9 with $\theta := \theta(k_1)$ and obtain $\theta_1, \theta_2 > 0$ such that if $g \in BC^1(\mathbb{R}, \mathbb{R})$ satisfies

(5.7.3) $$\|g'\|_{C^0} \leq \alpha + 1, \quad \|g - f\|_{C^0} \leq \theta_1, \text{ and}$$
$$\|(g - f)|\mathbb{R} \setminus (1/2 + (-\theta_2, \theta_2) + \mathbb{Z})\|_{C^1} \leq \theta_1,$$

then $\hat{g}$ satisfies conditions (ii, $\theta(k_1)$) and (4.8.1) of Proposition 4.8, i.e., $\hat{g}$ satisfies condition (5.7.2).

From 5.6, c), we have $\|-\alpha f'_\gamma\|_{C^0} \leq \alpha$ for all $\gamma > 0$, so the first inequality in (5.7.3) holds for all $\gamma > 0$. Properties 5.6 a), b) imply that there exists $\gamma_3 > 0$ such that for $\gamma \in (0, \gamma_3]$, the function $g = -\alpha f_\gamma$ satisfies also the remaining conditions in (5.7.3). Set $\gamma_1 := \min\{\gamma_2, \gamma_3\}$. For $\gamma \in (0, \gamma_1]$, $F_\gamma = -\alpha \hat{f}_\gamma$ satisfies conditions (5.7.1) and (5.7.2). □

6. Appendix (Auxiliary results)

Some statements of this section are slight variations of known results, adapted to a piecewise C^1 setting. Theorem 6.7 is an extension of Theorem 1.7 from [Lani-Wayda 3]. The statements of Lemma 6.8 on characteristic values, invariant eigenspaces and series expansion for solutions of $\dot{x}(t) = \alpha x(t-1)$ are special cases of well-known, considerably more general results. The last result, Proposition 6.9, shows that the numerical estimates needed in this work can be calculated by hand, independent of computers.

For the perturbation results of Section 4, we needed statements on piecewise continuous or piecewise C^1 equations, which are given below. We introduce some notation first.

Let $T > 0$. If there exist $n \in \mathbb{N}$, $t_1, ...t_n \in (0, T)$ with $t_1 < ... < t_n$, and if $A : [0, T] \setminus \{t_1, ..., t_n\} \to L_c(C, \mathbb{R})$ is continuous and such that the left and right limits

$$\lim_{t \to t_i-} A(t) \text{ and } \lim_{t \to t_i+} A(t) \text{ exist } (i = 1, ..., n),$$

then we say that A is *piecewise continuous* on $[0, T]$. The set of all such A is a vector space over $\mathbb{R}$ which we denote by $\mathcal{A}_T$. For $A \in \mathcal{A}_T$, we call the finitely many points where A is not defined the exceptional set of A, denoted by $E(A)$. On $\mathcal{A}_T$, we use two norms defined by

$$\|A\|_\infty := \sup\{|A(t)| \mid t \in [0, T] \setminus E(A)\},$$
$$\|A\|_1 := \int_0^T |A(s)| ds.$$

(The integral is the sum of the integrals over the subintervals where A is continuous.)

6.1. Definition and Remark. *Let $A \in \mathcal{A}_T$, $\psi \in C$. By a solution of the initial value problem*

$$(A, \psi) \qquad\qquad \dot{x}(t) = A(t)x_t, \quad x_0 = \psi$$

on $[-1, T]$ we mean a continuous function $x : [-1, T] \to \mathbb{R}$ with $x_0 = \psi$ such that $\dot{x}$ exists on $[0, T] \setminus E(A)$ (at 0 and T, we mean the one–sided derivative), and $\dot{x}(t) = A(t)x_t$ on $[0, T] \setminus E(A)$. For $A \in \mathcal{A}_T$ and $\psi \in C$, there exists a unique such solution which we denote by $x^{A, \psi}$.

Proof. (The equations considered here fall in the class of equations described in [Hale, Verduyn Lunel], p. 168.) □

6.2. Proposition. Let $T > 0$ and $A, B \in \mathcal{A}_T$, and $\varphi, \psi \in C$. Then, for $t \in [-1, T]$,

$$|x^{A,\psi}(t) - x^{B,\varphi}(t)| \leq [|\varphi - \psi| + |\varphi| \cdot \|A - B\|_1 \exp(\|B\|_1)] \exp(\|A\|_\infty T).$$

Proof. From linearity of the equation $\dot{x}(t) = A(t)x_t$ and from formula (1.4), p. 168 in [Hale, Verduyn Lunel], we get for $t \in [0, T]$

$$|x^{A,\psi}(t) - x^{A,\varphi}(t)| = |x^{A,\psi-\varphi}(t)| \leq |\psi - \varphi| \exp(t\|A\|_\infty).$$

It follows that one has

(6.2.1) $\qquad \forall t \in [-1, T] : |x^{A,\psi}(t) - x^{A,\varphi}(t)| \leq |\psi - \varphi| \exp(T\|A\|_\infty).$

Similarly,

(6.2.2) $\qquad \forall t \in [-1, T] : |x^{A,\varphi}(t)| \leq |\varphi| \exp(T\|A\|_\infty).$

Define now $\Delta(t) := x^{A,\varphi}(t) - x^{B,\varphi}(t)$ for $t \in [-1, T]$, and set

$$h(t) := (A(t) - B(t))x_t^{A,\varphi} \text{ for } t \in [0, T] \setminus (E(A) \cup E(B)).$$

Then $\Delta_0 = 0$, and for $t \in [0, T] \setminus (E(A) \cup E(B))$ we have

$$\dot{\Delta}(t) = (A(t) - B(t))x_t^{A,\varphi} + B(t)(x_t^{A,\varphi} - x_t^{B,\varphi}) =$$
$$= B(t)\Delta_t + h(t).$$

Using formula (1.4) from [Hale, Verduyn Lunel], p.168 again, it follows that for $t \in [0, T]$

$$|\Delta(t)| \leq [\int_0^t |h(s)|ds] \exp(\int_0^t |B(s)|ds).$$

From (6.2.2), we now get

$$|\Delta(t)| \leq |\varphi| \cdot \exp(T\|A\|_\infty) \cdot \int_0^t |A(s) - B(s)|ds \cdot \exp(\int_0^t |B(s)|ds,$$

so $|\Delta(t)| \leq |\varphi| \exp(T\|A\|_\infty) \cdot \|A - B\|_1 \cdot \exp(\|B\|_1)$. Combining the last estimate with (6.2.1) gives the assertion. $\square$

The author was not able to find a reference for the statement of the lemma below, which is an easy consequence of Gronwall's inequality.

Let $n \in \mathbb{N}$. For this lemma, we set $C := C^0([-1, 0], \mathbb{R}^n)$. Let $|\ |$ denote a norm on $\mathbb{R}^n$. The induced norm on C is also denoted by $|\ |$. For $\mathcal{U} \subset C$ open and $F \in BLip(\mathcal{U}, \mathbb{R}^n)$ and $\psi \in C$, there exists a unique maximal solution $y^{\psi,F} : [-1, t^+(\psi, F)) \to \mathbb{R}^n$ of the initial value problem $y_0 = \psi, \dot{y}(t) = F(y_t)$, where $t^+(\psi, F)$ is the supremum of the maximal existence interval. If $\mathcal{U} = C$ then $t_+(\psi, F) = \infty$.

6.3. Lemma. Let $T > 0$, and $\mathcal{U} \subset C$ open.

a) If $F, G \in BLip(\mathcal{U}, \mathbb{R}^n)$ and $y^{\psi,F}$, $y^{\varphi,G}$ are both defined on $[-1, T]$ at least, then one has for $t \in [-1, T]$

(6.3.1)
$$|y^{\psi,F}(t) - y^{\varphi,G}(t)| \leq (|\varphi - \psi| + T\|F - G\|_{C^0}) \exp(\min\{\mathrm{lip}(F), \mathrm{lip}(G)\}T).$$

The following map is continuous:
$$C \times (BLip(C, \mathbb{R}^n), \| \ \|_{C^0}) \to (C^0([-1, T], \mathbb{R}^n), \| \ \|_{C^0}),$$
$$(\psi, F) \mapsto y^{\psi,F}\big|_{[-1,T]}.$$

b) *The following map is also continuous:*
$$C \times (BLip(C, \mathbb{R}^n), \| \ \|_{C^0}) \to (C^1([0, T], \mathbb{R}^n), \| \ \|_{C^1}), \ \psi \mapsto y^{\psi,F}\big|_{[0,T]}.$$

c) *For fixed $F \in BLip(C, \mathbb{R}^n)$, the map*
$$C \to (C^1([0, T], \mathbb{R}^n), \| \ \|_{C^1}), \ \psi \mapsto y^{\psi,F}\big|_{[0,T]}$$

is Lipschitz continuous.

Proof. Ad a): Set $x := y^{\psi,F}$ and $y := y^{\varphi,G}$. For $s \in [-1, T]$, define $\xi(s) := \max_{\tau \in [-1,s]} |x(\tau) - y(\tau)|$. Then for $t \in [0, T]$

$$|x(t) - y(t)| \leq |\varphi - \psi| + \left| \int_0^t F(x_s) ds - \int_0^t G(y_s) ds \right|$$
$$\leq |\varphi - \psi| + \int_0^t |F(x_s) - F(y_s)| ds + \int_0^t |F(y_s) - G(y_s)| ds$$
$$\leq |\varphi - \psi| + \mathrm{lip}(F) \cdot \int_0^t |x_s - y_s| ds + \int_0^t \|F - G\|_{C^0} ds$$
$$\leq |\varphi + \psi| + T\|F - G\|_{C^0} + \mathrm{lip}(F) \cdot \int_0^t \xi(s) ds.$$

It follows that for $\tau \in [-1, t]$ one has

$$|x(\tau) - y(\tau)| \leq |\varphi - \psi| + T\|F - G\|_{C^0} + \mathrm{lip}(F) \int_0^t \xi(s) ds,$$

so $\xi(t) \leq |\varphi - \psi| + T\|F - G\|_{C^0} + \mathrm{lip}(F) \int_0^t \xi(s) ds$. Now Gronwall's inequality implies that, for $t \geq 0$,

$$\xi(t) \leq (|\varphi - \psi| + T\|F - G\|_{C^0}) \exp(\mathrm{lip}(F) \cdot t).$$

For all $t \in [-1, T]$, we thus have

$$|x(t) - y(t)| \leq (|\varphi - \psi| + T\|F - G\|_{C^0}) \exp((\text{lip} F) \cdot T).$$

It is clear that the corresponding estimate with $\text{lip}(G)$ instead of $\text{lip}(F)$ is also true, and thus (6.3.1) is proved. The continuity assertion of a) follows, if we apply (6.3.1) to the case $\mathcal{U} = C$. Further, for $t \in [0, T]$, we have

$$|\dot{x}(t) - \dot{y}(t)| = |F(x_t) - G(y_t)| \leq |F(x_t) - F(y_t)| + |F(y_t) - G(y_t)|$$
$$\leq \text{lip}(F)(|\varphi - \psi| + T\|F - G\|_{C^0}) \exp(\text{lip}(F)T) + \|F - G\|_{C^0}.$$

This estimate shows that for fixed $F \in BLip(C, \mathbb{R})$ one has

$$\|\dot{y}^{\psi, F} - \dot{y}^{\varphi, G}\|_{C^0([0,T], \mathbb{R})} \to 0 \text{ as } |\varphi - \psi| \to 0 \text{ and } \|G - F\|_{C^0} \to 0,$$

which proves assertion b). Assertion c) also follows from this estimate (set $G := F$). □

The following technical statement is used two times in the proof of Lemma 6.5 below.

6.4. Proposition. *Let $I \subset \mathbb{R}$ be an open interval, $n \in \mathbb{N}$, and let $\xi_1, .., \xi_n$ be n (not necessarily different) points in I. Set $X := \{\xi_1, .., \xi_n\}$ and assume that $f \in C^0(I, \mathbb{R})$ is BC^1 on $I \setminus X$, and the right and left limits $\lim_{x \to \xi \pm} f'(x)$ exist $(i = 1, .., n)$.*

Let $T > 0$, let $\tilde{I} \subset I$ be a compact interval, and let $Y \subset C^0([-1, T], \mathbb{R})$ be a family of functions with the following properties:
There exist $d > 0$ and $r_0 > 0$ such that, for all $y \in Y$,

(i) $y_{|[0,T]} \in C^1([0, T], \mathbb{R})$.
(ii) $y([-1, T]) \subset \tilde{I}$.
(iii) $\text{dist}(y([-1, 0]), X) \geq d$.
(iv) *If $t \in [1, T]$ and $y(t-1) = \xi_i \in X$ then for all $s \in [t-r_0, t+r_0] \cap [1, T]$ one has $|\dot{y}(s-1)| \geq d$, and $|y(s-1) - \xi| \geq d$ for all $\xi \in X \setminus \{\xi_i\}$.*
(v) *With $B_y := \bigcup\limits_{\substack{t \in [1,T], \\ y(t-1) \in X}} [t - r_0, t + r_0] \cap [1, T]$, one has*

$$\text{dist}[y(([1, T] \setminus B_y) - 1), X] \geq d.$$

Then for every $\varepsilon > 0$ there exists $\rho(\varepsilon) > 0$ such that for all $\delta \in C^0([-1, T], \mathbb{R})$ with the properties

(6.4.1) $\quad \delta_{|[0,T]} \in C^1, \ \|(\delta_{|[0,T]})'\|_{C^0} < d, \text{ and } \|\delta\|_{C^0} \leq \rho(\varepsilon),$

and all $y \in Y$, the following statements hold.

$y + \delta$ maps into I; there exist only finitely many points $t \in [0,T]$ with $y(t-1) + \delta(t-1) \in X$, and

$$\int_0^T |f'(y(s-1) + \delta(s-1)) - f'(y(s-1))| ds \leq \varepsilon.$$

Proof. Set $|f'| := \sup_{x \in I \setminus X} |f'(x)|$. For an arbitrary function $z : [-1,T] \to \mathbb{R}$, set $E(z) := \{t \in [0,T] \mid z(t-1) \in X\}$. For $y \in Y$, it follows from conditions (iii) and (iv) that for every $t \in E(y)$ one has $[t-r_0, t+r_0] \cap E(y) = \{t\}$. Thus, setting

$$m := \min\{k \in \mathbb{N} \mid k \geq (T-1)/r_0\},$$

we have

(6.4.2) $\qquad \forall y \in Y : \operatorname{card} E(y) \leq m.$

For $r \in (0, r_0]$ and $y \in Y$, set

$$B_{y,r} := \cup_{t \in E(y)} [t-r, t+r].$$

Conditions (iii), (iv) and (v) imply

(6.4.3) $\qquad \operatorname{dist}[y(([0,T] \setminus B_{y,r}) - 1), X] \geq \min\{d, dr\}.$

Let $\varepsilon > 0$. Set

$$r := \min\{\frac{\varepsilon}{8m(|f'|+1)}, 1/2, r_0\}, \text{ and } K := \tilde{I} \setminus \bigcup_{i=1}^n (\xi_i - dr, \xi_i + dr).$$

f' is uniformly continuous on the compact set K, which has positive distance to X. Hence there exists $\tilde{\rho}(\varepsilon) > 0$ such that

(6.4.4) $\qquad \forall a \in I, b \in K : |a-b| \leq \tilde{\rho}(\varepsilon) \implies |f'(b) - f'(a)| \leq \varepsilon/2T.$

There exists $\iota > 0$ with $\tilde{I} + [-\iota, \iota] \subset I$. Set $\rho(\varepsilon) := \min\{\tilde{\rho}(\varepsilon), dr/2, \iota\}$, and assume now that $\delta \in C^0([0,T], \mathbb{R})$ satisfies the conditions in (6.4.1) with this $\rho(\varepsilon)$. Take $y \in Y$. It follows from $\rho(\varepsilon) \leq \iota$ and property (ii) that $y + \delta$ maps into I. Since $\rho(\varepsilon) < dr < d$, property (iii) shows that

$$(y+\delta)([-1,0]) \cap X = \emptyset.$$

It follows from (6.4.3) and from $\rho(\varepsilon) \leq dr/2$ that

$$(y+\delta)[([0,T] \setminus B_{y,r}) - 1] \cap X = \emptyset.$$

For $s \in B_{y,r}$, property (iv) and (6.4.1) show that

$$|(y + \delta)(s - 1)| > d - d = 0.$$

Hence, using the last property in (iv), we see that for $t \in E(y)$ one has card $\{s \in [t - r, t + r] \mid (y + \delta)(s - 1) \in X\} \leq 1$. Together, we obtain

$$\text{card } E(y + \delta) \leq \text{card } E(y) \leq m.$$

Combined with the fact that $y + \delta$ maps into I, the last inequality shows that the integral in the assertion is defined. For $s \in [0, T] \setminus B_{y,r}$, the estimate (6.4.3), property (ii) and $r < 1$ imply $y(s - 1) \in K$. Hence (6.4.4) together with $|\delta|_{C^0} \leq \tilde{\rho}(\varepsilon)$ shows that

$$\forall s \in [0, T] \setminus B_{y,r} : |f'(y(s - 1) + \delta(s - 1)) - f'(y(s - 1))| \leq \varepsilon/2T.$$

Using (6.4.2), we get the estimate

$$\int_{B_{y,r}} |f'(y(s - 1) + \delta(s - 1)) - f'(y(s - 1))| ds \leq \sum_{t \in E(y)} 2r \cdot 2|f'| \leq 4mr|f'|.$$

Combining the last two estimates with the choice of r, we get

$$\int_0^T |..| ds \leq \int_{[0,T] \setminus B_{y,r}} ... + \int_{B_{y,r}} ... \leq T \cdot \varepsilon/2T + \frac{\varepsilon 4m|f'|}{8m(|f'| + 1)} \leq \varepsilon. \quad \square$$

6.5. Lemma. Let $f \in BLip(\mathbb{R}, \mathbb{R})$, and let $x : [-1, \infty) \to \mathbb{R}$ be a solution of (f) $\dot{x}(t) = f(x(t - 1))$. Let $T > 1$, and let $I \subset \mathbb{R}$ be an open interval with

$$x([-1, T]) \subset I.$$

Assume that there exist $n \in \mathbb{N}$ and $\tau_1, ..., \tau_n \in (1, T)$, $\tau_1 < ... < \tau_n$ such that with $\xi_i := x(\tau_i - 1)$ $(i = 1, ..., n)$, the following statements hold.

1) f is continuously differentiable on $I \setminus \{\xi_1, ..., \xi_n\}$, and the right and left limits of f' exist at these points.
2) For $t \in [0, T]$, one has the equivalence

$$x(t - 1) \in \{\xi_1, ..., \xi_n\} \iff t \in \{\tau_1, ..., \tau_n\}.$$

3) $\dot{x}(\tau_i - 1) \neq 0$, $i = 1, ..., n$.

Then there exists a neighborhood U of x_0 in C with the following properties:

a) The set $Y := \{y^\psi|_{[-1,T]} \mid \psi \in U\}$ satisfies the conditions of Proposition 6.4. (Here y^ψ is the solution of equation (f) with initial value ψ.)
b) $D_2 \Phi_f(T, \cdot)$ exists and is uniformly continuous on U.

c) For $\psi \in U$ and $\chi \in C$, $D_2\Phi_f(T,\psi)\chi$ is given by $w_T^{\psi,\chi}$, where $w^{\psi,\chi}:$ $[-1,\infty) \to \mathbb{R}$ is the solution of

$$w_0 = \chi, \quad \dot{w}(t) = f'(y^\psi(t-1))w(t-1) \quad (t \in [0,T])$$

in the sense of Definition and Remark 6.1 (with $A(t)\varphi := f'(x(t-1))\varphi(-1)$ for $t \in [0,T] \setminus \{\tau_1,...,\tau_n\}$, $\varphi \in C$).

Proof. Set $d_1 = \min\{|\dot{x}(\tau_i - 1)| \mid i \in \{1,....,n\}\}$, and set $\lambda := \min\{|\xi_i - \xi_j| \mid i,j \in \{1,...,n\}, \xi_i \neq \xi_j\}$ if there are at least two different numbers in the set

$$X := \{\xi_1,...,\xi_n\},$$

and set $\lambda := 1$ otherwise. Choose a compact interval $\tilde{I} \subset I$ such that $x([-1,T]) \subset \text{int}(\tilde{I})$. There exists $r_0 > 0$ such that with $B_2 := \bigcup_{i=1}^n [\tau_i - 2r_0, \tau_i + 2r_0]$ and $B_1 := \bigcup_{i=1}^n [\tau_i - r_0, \tau_i + r_0]$, the subsequent statements hold.

$$[\tau_i - 2r_0, \tau_i + 2r_0] \cap [\tau_j - 2r_0, \tau_j + 2r_0] = \emptyset \text{ if } i,j \in \{1,...,n\}, i \neq j.$$
$$2r_0\|f\|_{C^0} \leq \lambda/2.$$
$$B_2 \subset (1,T).$$
$$\forall t \in B_2 : |\dot{x}(t-1)| \geq d_1/2.$$
$$\text{dist}[x(([0,T] \setminus B_1) - 1), X] > 0.$$

It follows from equation (f) and the choice of r_0 that for $i \in \{1,...,n\}$ and $t \in [\tau_i - 2r_0, \tau_i + 2r_0]$ one has $|x(t-1) - \xi_i| \leq \|f\|_{C^0} \cdot 2r_0 \leq \lambda/2$, which implies

(6.5.1) $\qquad |x(t-1) - \xi| \geq \lambda/2$ for all $\xi \in X \setminus \{\xi_i\}$.

Using Lemma 6.3, c), and (6.3.1), we see that there exist a neighborhood U of x_0 in C and $d_2 > 0$ such that for all $\psi \in U$ the following statements are true.

(6.5.2) $\qquad \forall t \in B_2 : |\dot{y}^\psi(t-1)| \geq d_1/4.$

(6.5.3) $\qquad \text{dist}[y^\psi(([0,T] \setminus B_1) - 1), X] \geq d_2.$

(6.5.4) $\qquad y^\psi([-1,T]) \subset \tilde{I}.$

(6.5.5) $\quad \forall i \in \{1,...,n\} \, \forall t \in [\tau_i - 2r_0, \tau_i + 2r_0] \, \forall \xi \in X \setminus \{\xi_i\} :$
$$|y^\psi(t-1) - \xi| \geq \lambda/4.$$

Proof of assertion a): With Y defined as in assertion a), using U from above, it is clear that property (i) from Proposition 6.5 holds. Property (ii)

holds in view of (6.5.4). We verify properties (iii)-(v): It follows from (6.5.3) and from $B_1 \subset B_2 \subset (1, T)$ that, for $\psi \in U$,

(6.5.6) $$\operatorname{dist}(y^\psi([-1, 0]), X) \geq d_2.$$

If $\psi \in U$, $t \in [1, T]$ and $y^\psi(t - 1) \in X$ then $t \in B_1$ (see (6.5.3)), so there exists a unique $i \in \{1, ..., n\}$ with $[t - r_0, t + r_0] \subset [\tau_i - 2r_0, \tau_i + 2r_0] \subset B_2$, and we obtain from (6.5.2) and (6.5.5) that

(6.5.7) $$|\dot{y}^\psi(s - 1)| \geq d_1/4 \text{ for } s \in [t - r_0, t + r_0],$$
$$\forall \xi \in X \setminus \{\xi_i\} : |y^\psi(s - 1) - \xi| \geq \lambda/4.$$

Let now $\psi \in U$, set $y := y^\psi$, and assume that, with B_y defined as in property (v) from Proposition 6.4, $s \in [1, T] \setminus B_y$. If $s \in [1, T] \setminus B_1$ then, from (6.5.3), $\operatorname{dist}(y(s - 1), X) \geq d_2$.

Otherwise, $s \in B_1$, so there exists a unique $i \in \{1, ..., n\}$ with $s \in [\tau_i - r_0, \tau_i + r_0]$. From (6.5.2), we have $|\dot{y}^\psi(\cdot - 1)| \geq d_1/4$ on $[\tau_i - r_0, \tau_i + r_0]$.

First case: $y(t - 1) \notin X$ for all $t \in [\tau_i - r_0, \tau_i + r_0]$. Then

$$\operatorname{dist}(y(s - 1), X) \geq \min\{\operatorname{dist}[y(\tau_i - r_0 - 1), X], \operatorname{dist}[y(\tau_i + r_0 - 1), X]\}.$$

Since $\{\tau_i - r_0, \tau_i + r_0\} \subset \operatorname{clos}([0, T] \setminus B_1)$, we conclude from (6.5.3)

(6.5.8) $$\operatorname{dist}(y(s - 1), X) \geq d_2.$$

Second case: There exists $t \in [\tau_i - r_0, \tau_i + r_0]$ with $y(t - 1) \in X$. Then we see from (6.5.5) that $y(t - 1) = \xi_i$. Now $s \notin B_y$ implies $|s - t| \geq r_0$, so using (6.5.2) we get $|y(s - 1) - \xi_i| \geq r_0 d_1/4$. Together with (6.5.5), we obtain in the second case

(6.5.9) $$\operatorname{dist}(y(s - 1), X) \geq \min\{\lambda/4, r_0 d_1/4\}.$$

Set now $d := \min\{d_2, d_1/4, \lambda/4, r_0 d_1/4\}$. Property (iii) then follows from (6.5.6), property (iv) from (6.5.7), and property (v) from (6.5.8) and (6.5.9). Assertion a) is proved.

Proof of assertions b) and c): For $\psi, \chi \in C$, set

$$\delta^{\psi,\chi}(t) := y^{\psi+\chi}(t) - y^\psi(t) \quad (t \in [-1, \infty)).$$

It follows from Lemma 6.3, c) that there exists $L > 0$ such that

(6.5.10) $$\forall \psi, \chi \in C \; \forall t \in [-1, T] : |\delta^{\psi,\chi}(t)| \leq L|\chi|,$$
$$\forall t \in [0, T] : |\dot{\delta}^{\psi,\chi}(t)| \leq L|\chi|.$$

For $\psi \in U$, part a) of the present lemma and condition (iv) of Proposition 6.4 show that $f'(y^\psi(t-1))$ is defined for all $t \in [0, T]$ except for finitely many, at which the right and left limits exist. Thus the solution $w^{\psi,\chi}$ from assertion c) is defined in the sense of 6.1 (for all $\chi \in C$), and we can set

$$\Delta^{\psi,\chi}(t) := \delta^{\psi,\chi}(t) - w^{\psi,\chi}(t) \quad (t \in [-1, T]).$$

Fix now $\psi \in U$. To prove existence of $D_2 \Phi_f(T, \psi)$ and assertion c), we have to show that for every $\varepsilon > 0$ there exists $\delta > 0$ such that $|\chi| < \delta$ implies

(6.5.11) $$|\Delta_T^{\psi,\chi}| \leq \varepsilon |\chi|.$$

For $\chi \in C$ such that $\psi + \chi \in U$, we have for all $t \in [0, T]$, except for finitely many,

$$\dot{\Delta}^{\psi,\chi}(t) =$$
$$= f(y^{\psi+\chi}(t-1)) - f(y^\psi(t-1) - f'(y^\psi(t-1))w^{\psi,\chi}(t-1)$$
$$= \int_0^1 f'(y^\psi(t-1) + s \cdot \delta^{\psi,\chi}(t-1))ds\, \delta^{\psi,\chi}(t-1) - $$
$$\quad - f'(y^\psi(t-1))w^{\psi,\chi}(t-1)$$
$$= \int_0^1 [f'(y^\psi(t-1) + s \cdot \delta^{\psi,\chi}(t-1)) - f'(y^\psi(t-1))]ds\, \delta^{\psi,\chi}(t-1) + $$
$$\quad + f'(y^\psi(t-1))[\delta^{\psi,\chi}(t-1) - w^{\psi,\chi}(t-1)]$$
$$= \int_0^1 [f'(y^\psi(t-1) + s \cdot \delta^{\psi,\chi}(t-1)) - f'(y^\psi(t-1))]ds\, \delta^{\psi,\chi}(t-1) + $$
$$\quad + f'(y^\psi(t-1))\Delta^{\psi,\chi}(t-1).$$

(Note that we have used the Fundamental Theorem of Calculus for the continuous and piecewise C^1 function $f_{|I}$.) For $t \in [0, T]$ as above, set

$$h^{\psi,\chi}(t) := \int_0^1 [f'(y^\psi(t-1) + s \cdot \delta^{\psi,\chi}(t-1)) - f'(y^\psi(t-1))]ds\, \delta^{\psi,\chi}(t-1).$$

The function $h^{\psi,\chi}$ is continuous at all these t, since

$$h^{\psi,\chi}(t) = f(y^\psi(t-1) + \delta^{\psi,\chi}(t-1)) - f(y^\psi(t-1)) - f'(y^\psi(t-1))\delta^{\psi,\chi}(t-1).$$

It follows from formula 1.4 on p. 168 of [Hale, Verduyn Lunel] that, for all $t \in [0, T]$,

$$|\Delta^{\psi,\chi}(t)| \leq \int_0^t |h^{\psi,\chi}(s)|ds \exp[\int_0^t |f'(y^\psi(s-1))|ds].$$

APPENDIX (AUXILIARY RESULTS)

Set $|f'| := \sup\{|f'(x)| \mid x \in I \setminus \{\xi_1, ..., \xi_n\}\}$, and set $E := \exp(T|f'|)$. Then the last estimate implies

(6.5.12) $$|\Delta_T^{\psi,\chi}| \leq \int_0^T |h^{\psi,\chi}(s)| ds \cdot E.$$

Now

$$\int_0^T |h^{\psi,\chi}(s)| ds \leq$$
$$\leq \int_0^1 \int_0^T |f'(y^\psi(t-1) + s \cdot \delta^{\psi,\chi}(t-1)) - f'(y^\psi(t-1))| |\delta^{\psi,\chi}(t-1)| dt \cdot ds$$
$$\leq \int_0^1 \int_0^T |f'(y^\psi(t-1) + s \cdot \delta^{\psi,\chi}(t-1)) - f'(y^\psi(t-1))| dt \cdot ds \|\delta^{\psi,\chi}\|_{C^0}.$$

Let $\varepsilon > 0$. Recall $d > 0$ from the proof of part a). It follows from part a) and Proposition 6.4 that there exists $\rho_1 > 0$ (dependent on ε) such that if

(6.5.13) $$\|(\delta^{\psi,\chi}|_{[0,T]})\|_{C^0} < d \text{ and } \|\delta^{\psi,\chi}\|_{C^0} \leq \rho_1$$

then, for all $s \in [0,1]$,

(6.5.14) $$\int_0^T |f'(y^\psi(t-1) + s \cdot \delta^{\psi,\chi}(t-1)) - f'(y^\psi(t-1))| dt \leq \varepsilon/EL.$$

From (6.5.10) we see that if $|\chi| < \min\{\operatorname{dist}(\psi, \partial U), d/L, \rho_1/L\}$, then (6.5.13) and, hence, (6.5.14) holds. Combining this estimate with (6.5.12) and (6.5.10), we get for such χ

$$|\Delta_T^{\psi,\chi}| \leq \frac{\varepsilon}{EL} L|\chi|E = \varepsilon|\chi|,$$

so (6.5.11) holds. The assertions of b) and c) are proved, except for uniform continuity of $D_2\Phi_f(T,\cdot)$ on U.

Let $\varepsilon > 0$. Applying Proposition 6.4 and (6.5.10) again, we obtain that there exists $\rho_2 > 0$ (dependent on ε) such that if $\varphi, \psi \in U$ and $|\varphi - \psi| < \min\{d/L, \rho_2/L\}$ then

$$\|f'(y^\varphi(\cdot - 1)) - f'(y^\psi(\cdot - 1))\|_1$$
$$= \int_0^T |f'(y^\varphi(t-1)) - f'(y^\psi(t-1))| dt \leq \varepsilon/\exp(2|f'|T).$$

It follows now from the already proved statement c) and from Proposition 6.2 that, for $\chi \in C$,

$$|D_2\Phi_f(T,\varphi)\chi - D_2\Phi_f(T,\psi)\chi| = |w_T^{\varphi,\chi} - w_T^{\psi,\chi}|$$
$$\leq |\chi| \cdot \|f'(y^\varphi(\cdot-1)) - f'(y^\psi(\cdot-1))\|_1 \exp(\|f'(y^\psi(\cdot-1))\|_1)\exp(|f'|T)$$
$$\leq |\chi| \cdot \frac{\varepsilon}{\exp(2|f'|T)} \exp(|f'|T)\exp(|f'|T)$$
$$= \varepsilon|\chi|. \qquad \square$$

The following result sometimes helps to avoid the repetition of similar, triangle-inequality-type arguments. Unfortunately, such statements are not a standard content of analysis books.

6.6. Proposition (Composition of BC^1 maps). *Let $n \in \mathbb{N}$ and normed spaces E_i, $(i = 1, ..., n+1)$, and open subsets $U_i \subset E_i$ $(i = 1, ..., n)$ be given. Assume that Λ is a topological space, and that, for $\lambda \in \Lambda$, maps $f_{i,\lambda} \in BC^1(U_i, E_{i+1})$ are given such that*

$$\forall \lambda \in \Lambda: f_{i,\lambda}(U_i) \subset U_{i+1} \quad (i=1,...,n).$$

Let $\lambda_0 \in \Lambda$ and assume the following conditions.
a) $\|f_{i,\lambda} - f_{i,\lambda_0}\|_{C^1} \to 0$ as $\lambda \to \lambda_0$, for $i = 1, ..., n$.
b) The maps f_{i,λ_0} and Df_{i,λ_0} are uniformly continuous on U_i for $i = 2, ..., n$, if $n \geq 2$.

Then the maps $F_\lambda : U_1 \to E_{n+1}$, $F_\lambda := f_{n,\lambda} \circ ... \circ f_{1,\lambda}$ satisfy $F_\lambda \in BC^1(U_1, E_{n+1})$ $(\lambda \in \Lambda)$, and

$$\|F_\lambda - F_{\lambda_0}\|_{C^1} \to 0 \text{ as } \lambda \to \lambda_0.$$

Further, if also f_{1,λ_0} and Df_{1,λ_0} are uniformly continuous, then F_{λ_0} and DF_{λ_0} are uniformly continuous.

Proof. (Induction on n.) For $n = 1$, the assertion is trivial. Let $n \geq 2$, assume that the assertion is true for $n - 1 \in \mathbb{N}$, and let $f_{i,\lambda}$ and U_i $(i = 1, ..., n)$ as in the statement of the proposition be given. From the induction hypotheses, we know that, for $\lambda \in \Lambda$,

(6.6.1) $\qquad G_\lambda := f_{n-1,\lambda} \circ ... \circ f_{1,\lambda} \in BC^1(U_1, E_n)$, and

(6.6.2) $\qquad \|G_\lambda - G_{\lambda_0}\|_{C^1} \to 0$ as $\lambda \to \lambda_0$.

It follows from the chain rule and (6.6.1), together with $f_{n,\lambda} \in BC^1$, that $F_\lambda \in BC^1(U_1, E_{n+1})$ $(\lambda \in \Lambda)$. We denote the norms on E_i and on $L_c(E_i, E_{i+1})$ by $|\ |$ $(i = 1, ..., n)$.

APPENDIX (AUXILIARY RESULTS)

Let $\varepsilon > 0$. Because of assumption b) and $n \geq 2$, there exists $\delta > 0$ such that
(6.6.3)
$$\forall x, y \in U_n, |x - y| \leq \delta : |f_{n,\lambda_0}(x) - f_{n,\lambda_0}(y)| \leq \varepsilon/2,$$
$$|Df_{n,\lambda_0}(x) - Df_{n,\lambda_0}(y)| \leq \frac{\varepsilon}{3(\|DG_{\lambda_0}\|_{C^0} + 1)}.$$

From (6.6.2) and assumption a), there exists a neighborhood Λ_1 of λ_0 in Λ such that for all $\lambda \in \Lambda_1$

(6.6.4)
$$\|G_\lambda - G_{\lambda_0}\|_{C^1} \leq \min\{1, \delta, \frac{\varepsilon}{3(\|Df_{n,\lambda_0}\|_{C^0} + 1)}\},$$
$$\|f_{n,\lambda} - f_{n,\lambda_0}\|_{C^0} \leq \varepsilon/2,$$
$$\|Df_{n,\lambda} - Df_{n,\lambda_0}\|_{C^0} \leq \frac{\varepsilon}{3(\|DG_{\lambda_0}\|_{C^0} + 1)}.$$

Note that, for $\lambda \in \Lambda_1$,

(6.6.5)
$$\|DG_\lambda\|_{C^0} \leq 1 + \|DG_{\lambda_0}\|_{C^0}.$$

It follows that for $\lambda \in \Lambda_1$ and $x \in U_1$,

$$|F_\lambda(x) - F_{\lambda_0}(x)| = |f_{n,\lambda}(G_\lambda(x)) - f_{n,\lambda_0}(G_{\lambda_0}(x))|$$
$$\leq \|f_{n,\lambda} - f_{n,\lambda_0}\|_{C^0} + |f_{n,\lambda_0}(G_\lambda(x)) - f_{n,\lambda_0}(G_{\lambda_0}(x))|$$
$$\leq \varepsilon/2 + \varepsilon/2 = \varepsilon,$$

where we used the first inequality of (6.6.3), and the first two of (6.6.4). Further, combining (6.6.3)-(6.6.5), we get

$$|DF_\lambda(x) - DF_{\lambda_0}(x)|$$
$$= |Df_{n,\lambda}(G_\lambda(x))DG_\lambda(x) - Df_{n,\lambda_0}(G_{\lambda_0}(x))DG_{\lambda_0}(x)|$$
$$\leq \{|Df_{n,\lambda}(G_\lambda(x)) - Df_{n,\lambda_0}(G_\lambda(x))| + |Df_{n,\lambda_0}(G_\lambda(x)) - Df_{n,\lambda_0}(G_{\lambda_0}(x))|\} \cdot$$
$$\cdot |DG_\lambda(x)| + |Df_{n,\lambda_0}(G_{\lambda_0}(x))| \cdot |DG_\lambda(x) - DG_{\lambda_0}(x)|$$
$$\leq \{\|Df_{n,\lambda} - Df_{n,\lambda_0}\|_{C^0} + \frac{\varepsilon}{3(\|DG_{\lambda_0}\|_{C^0} + 1)}\} \cdot (\|DG_{\lambda_0}\|_{C^0} + 1) +$$
$$+ \|Df_{n,\lambda_0}\|_{C^0} \cdot \frac{\varepsilon}{3(\|Df_{n,\lambda_0}\|_{C^0} + 1)}$$
$$\leq \varepsilon/3 + \varepsilon/3 + \varepsilon/3 = \varepsilon.$$

This proves the asserted convergence. Assume now that also f_{1,λ_0} and Df_{1,λ_0} are uniformly continuous. We know then, from the induction hypotheses, that G_{λ_0} and DG_{λ_0} are uniformly continuous. It follows from uniform continuity of f_{n,λ_0} and from

$$F_{\lambda_0}(x) - F_{\lambda_0}(y) = f_{n,\lambda_0}(G_{\lambda_0}(x)) - f_{n,\lambda_0}(G_{\lambda_0}(y)) \quad (x, y \in U_1)$$

that F_{λ_0} is uniformly continuous. Further, since

$$DF_{\lambda_0}(x) - DF_{\lambda_0}(y) = Df_{n,\lambda_0}(G_{\lambda_0}(x))DG_{\lambda_0}(x) - Df_{n,\lambda_0}(G_{\lambda_0}(y))DG_{\lambda_0}(y)),$$

uniform continuity of DF_{λ_0} follows from uniform continuity of G_{λ_0}, and from uniform continuity and boundedness of Df_{n,λ_0} and DG_{λ_0}. □

Theorem 6.7 below is a stronger version of Theorem 1.7 from [Lani-Wayda 3]. The difference is that a Poincaré-type map is constructed not only in the neighborhood of a point, but of a possibly larger set, and that the functional defining the surface of section is allowed to be nonlinear. For Theorem 6.7, we set $C := C^0([-1,0], \mathbb{R}^n)$, where $n \in \mathbb{N}$. For $\mathcal{U} \subset C$ open and $F \in BC^1 Lip(\mathcal{U}, \mathbb{R}^n)$, there exists a local semiflow $\Phi_F : \bigcup_{\psi \in \mathcal{U}} [0, t^+(\psi, F)) \times \{\psi\} \to \mathcal{U} \subset C$, where $t^+(\psi, F)$ is the supremum of the maximal existence interval of the solution of (F) through ψ. (If $\mathcal{U} = C$ then $t^+(\psi, F) = \infty$ for all $\psi \in C$.)

6.7. Theorem. *Let $\mathcal{U} \subset C$ and $\mathcal{V} \subset \mathcal{U}$ both be open. Assume that $F \in BC^1 Lip(\mathcal{U}, \mathbb{R}^n)$ and $h \in BC^1 Lip(\mathcal{V}, \mathbb{R}^n)$. Let $D \subset \mathcal{U}$ be bounded, and assume that a uniformly continuous map $\tau_0 : D \to \mathbb{R}$ and numbers $\theta_0, T, d_0 > 0$ are given such that the following four properties hold for all $\varphi \in D$.*

(i) $1 < \tau_0(\varphi) - \theta_0 < \tau_0(\varphi) + \theta_0 < t^+(\varphi, F)$.
(ii) $\tau_0(\varphi) + \theta_0 \leq T$.
(iii) $\Phi_F(\tau_0(\varphi), \varphi) \in \mathcal{V}$, and $h(\Phi_F(\tau_0(\varphi), \varphi)) = 0$.
(iv) $|\frac{d}{dt}[t \mapsto h(\Phi_F(t, \varphi))]|_{t = \tau_0(\varphi)}| \geq d_0$.

Further, we assume

(v) $\mathrm{dist}(\{\Phi_F(t, \varphi) \mid \varphi \in D, t \in [0, \tau_0(\varphi) + \theta_0]\}, \partial \mathcal{U}) > 0$.
(vi) $\mathrm{dist}(\{\Phi_F(\tau_0(\varphi), \varphi) \mid \varphi \in D\}, \partial \mathcal{V}) > 0$.

Then there exist $r > 0$, $\beta > 0$, $\theta \in (0, 1)$ such that, with the sets

$$W := \{\psi \in \mathcal{U} \mid \mathrm{dist}(\psi, D) < r\},$$

$$\mathcal{B} := \{G \in BC^1 Lip(\mathcal{U}, \mathbb{R}^n) \mid \|G - F\|_{C^1} < \beta\},$$

the following statements hold.

1) *There exists a BC^1 map $\tau : W \times \mathcal{B} \to \mathbb{R}$ with the properties below.*

$$\forall \varphi \in D : \tau(\varphi, F) = \tau_0(\varphi).$$

For $\chi \in W$, $G \in \mathcal{B}$, one has $\tau(\chi, G) > 1$, and $\Phi_G(\cdot, \chi)$ is defined on $[0, \tau(\chi, G) + \theta]$, and

$$\forall t \in \tau(\chi, G) + (-\theta, \theta) : \quad h(\Phi_G(t, \chi)) = 0 \iff t = \tau(\chi, G).$$

2) *The maps $P_G: W \to C$, $P_G(\chi) := \Phi_G(\tau(\chi, G), \chi)$ are in $BC^1(W, C)$. If $\tilde{G} \in \mathcal{B}$ and $D\tilde{G}$ is uniformly continuous, then the map*

$$(\mathcal{B}, \|\ \|_{C^1}) \ni G \mapsto P_G \in (BC^1(W, C), \|\ \|_{C^1})$$

is continuous at $\tilde{G}$, and $DP_{\tilde{G}}$ is uniformly continuous.

Proof. Set $\Omega := \bigcup_{\substack{\psi \in \mathcal{U}, \\ G \in BC^1 Lip(\mathcal{U}, \mathbb{R}^n)}} [0, t^+(\psi, G)) \times \{\psi\} \times \{G\}$. Similar to the case of ordinary differential equations, Ω is open in $[0, \infty) \times C \times BC^1 Lip(\mathcal{U}, \mathbb{R}^n)$. It follows from Lemma 1.5 in [Lani-Wayda 3] that the map

$$\Omega \ni (t, \psi, G) \mapsto \Phi_G(t, \psi) \in C$$

is C^1 on the set $\{(t, \psi, G) \in \Omega \mid t > 1\}$. Nonlinearities $G \in BC^1 Lip(\mathcal{U}, \mathbb{R}^n)$ can be extended to Lipschitz functions $\tilde{G} \in Lip(C, \mathbb{R}^n)$. The semiflows $\Phi_{\tilde{G}}$ are then defined on $\mathbb{R}_0^+ \times C$.

Properties (v) and (ii), together with estimate (6.3.1) from Lemma 6.3 show that there exist bounded neighborhoods U_0 of 0 in C and $\mathcal{B}_0$ of F in $BC^1 Lip(\mathcal{U}, \mathbb{R}^n)$ such that $\Phi_{\tilde{G}}(t, \varphi + \psi) \in \mathcal{U}$ for $\varphi \in D$, $t \in [0, \tau_0(\varphi) + \theta_0]$, and for $G \in \mathcal{B}_0$, independent of the extension $\tilde{G}$. Hence we have

(6.7.1) $\qquad \forall G \in \mathcal{B}_0 \, \forall \varphi \in D \, \forall \psi \in U_0 : \tau_0(\varphi) + \theta_0 < t^+(\varphi + \psi, G),$

so that $\Phi_G(t, \varphi + \psi)$ is defined and an element of $\mathcal{U}$ for such φ, ψ and for $t \in [0, \tau_0(\varphi) + \theta_0]$. Set

$$\mathcal{M} := \bigcup_{\varphi \in D, \psi \in U_0, G \in \mathcal{B}_0} (1, \tau_0(\varphi) + \theta_0] \times \{\varphi + \psi\} \times \{G\}.$$

(6.7.2) *Claim:* The map $\Phi : \mathcal{M} \ni (t, \varphi + \psi, G) \mapsto \Phi_G(t, \varphi + \psi)$ is BC^1.

Proof: We know that this map is C^1. Boundedness of Φ follows from boundedness of U_0 and $\mathcal{B}_0$ and from estimate (6.3.1). It remains to obtain bounds on the partial derivatives.

1. For $(t, \chi, G) \in \mathcal{M}$, $D_1 \Phi_G(t, \chi)(1)$ is given by the segment at time t of the function $s \mapsto G(y_s^{\chi, G})$, which segment we denote by $G(y_\cdot^{\chi, G})_t$. Thus boundedness of $\mathcal{B}_0$ implies boundedness of $D_1 \Phi$.

2. For $(t, \chi, G) \in \mathcal{M}$ and $\eta \in C$, the value of $D_2 \Phi_G(t, \chi) \eta$ is given by $w_t^{\chi, \eta, G}$, where $w^{\chi, \eta, G}$ is the solution of $w_0 = \eta$, $\dot{w}(s) = DG(y_s^{\chi, G}) w_s$ ($s \geq 0$) (compare [Hale, Verduyn Lunel], p. 49, Thm. 4.1). Since $t \leq T$ (from property (ii)), boundedness of $D_2 \Phi$ follows from boundedness of $\mathcal{B}_0$, in view of estimate (6.2.2) (from the proof of Proposition 6.2).

3. (We denote the derivative of Φ with respect to the argument $G \in \mathcal{B}_0$ by $D_3 \Phi$.) For $(t, \chi, G) \in \mathcal{M}$ and $H \in BC^1 Lip(\mathcal{U}, \mathbb{R}^n)$, one has

$$D_3 \Phi_G(t, \chi) H = \left[\int_0^\cdot DG(y_s^{\chi, G}) D_3 \Phi_G(s, \chi) H \, ds + \int_0^\cdot H(y_s^{\chi, G}) ds \right]_t.$$

(This formula follows, e.g., from the integral equation on p. 16 of [Lani-Wayda 3].) Similar to the proof of Proposition 6.2, we obtain from this formula, and from $t \leq T$, the bound

$$|D_3 \Phi_G(t, \mathcal{X}) H| \leq T \|H\|_{C^0} \exp(T \|DG\|_{C^0}).$$

Boundedness of $\mathcal{B}_0$ now shows that $D_3 \Phi_G$ is bounded. Claim (6.7.2) is proved.

We prepare an estimate that expresses continuity of Φ. Let $(t_i, \mathcal{X}_i, G_i) \in \mathcal{M}$ $(i = 1, 2)$. If $t_1 \leq t_2$ then, using (6.3.1) and equation (G_2), we get

$$|y_{t_1}^{\mathcal{X}_1, G_1} - y_{t_2}^{\mathcal{X}_2, G_2}|$$
$$\leq |y_{t_1}^{\mathcal{X}_1, G_1} - y_{t_1}^{\mathcal{X}_2, G_2}| + |y_{t_1}^{\mathcal{X}_2, G_2} - y_{t_2}^{\mathcal{X}_2, G_2}|$$
$$\leq (|\mathcal{X}_1 - \mathcal{X}_2| + t_1 \|G_1 - G_2\|_{C^0}) \exp[\min\{\text{lip}(G_1), \text{lip}(G_2)\} t_1] +$$
$$+ \|G_2\|_{C^0} (t_2 - t_1)$$
$$\leq (|\mathcal{X}_1 - \mathcal{X}_2| + T \|G_1 - G_2\|_{C^0}) \exp[\min\{\text{lip}(G_1), \text{lip}(G_2)\} T] +$$
$$+ \max\{\|G_1\|_{C^0}, \|G_2\|_{C^0}\} |t_2 - t_1|.$$

We abbreviate the last expression with $\Delta(\mathcal{X}_1, \mathcal{X}_2, G_1, G_2, t_1, t_2)$. Since it is symmetric with respect to the indices 1 and 2, it follows that also in case $t_1 > t_2$ we have

(6.7.3) $$|y_{t_1}^{\mathcal{X}_1, G_1} - y_{t_2}^{\mathcal{X}_2, G_2}| \leq \Delta(\mathcal{X}_1, \mathcal{X}_2, G_1, G_2, t_1, t_2).$$

Using this estimate, and properties (i) and (vi), one sees that there exist open neighborhoods $\mathcal{B}_1 \subset \mathcal{B}_0$ of F in $BC^1 Lip(\mathcal{U}, \mathbb{R}^n)$ and $U_1 \subset U_0$ of 0 in C, and $\theta_1 \in (0, \theta_0]$ such that with

$$\mathcal{N} := \{(t, \varphi + \psi, G) \mid \varphi \in D, \psi \in U_1, t \in \tau_0(\varphi) + (-\theta_1, \theta_1), G \in \mathcal{B}_1\},$$

one has $\mathcal{N} \subset \mathcal{M}$, and

(6.7.4) $$\Phi(\mathcal{N}) \subset \mathcal{V}, \text{ and } \text{dist}(\Phi(\mathcal{N}), \partial \mathcal{V}) > 0.$$

(Note that $\mathcal{N}$ is an open subset of $(1, T] \times C \times BC^1 Lip(\mathcal{U}, \mathbb{R}^n)$.) Define

$$\Pi : \mathcal{N} \to \mathbb{R}, \quad \Pi(t, \mathcal{X}, G) := h(\Phi_G(t, \mathcal{X})).$$

Π is C^1. From properties (iii) and (iv), we have for $\varphi \in D$

(6.7.5) $$\Pi(\tau_0(\varphi), \varphi, F) = 0, \quad |D_1 \Pi(\tau_0(\varphi), \varphi, F)| \geq d_0.$$

APPENDIX (AUXILIARY RESULTS)

Applying the Implicit Function Theorem, we see that for every $\varphi \in D$ there exist positive numbers $r_\varphi, \theta_\varphi$ and β_φ such that

$$C(r_\varphi) \subset U_1, \ \mathcal{B}_\varphi := \{G \in BC^1 Lip(\mathcal{U}, \mathbb{R}^n) \mid \|G - F\|_{C^1} < \beta_\varphi\} \subset \mathcal{B}_1,$$

and a BC^1 map

$$\tau_\varphi : (\varphi + C(r_\varphi)) \times \mathcal{B}_\varphi \to I_\varphi := (\tau_0(\varphi) - \theta_\varphi, \tau_0(\varphi) + \theta_\varphi) \subset \mathbb{R}$$

such that
(6.7.6)
$$\forall \psi \in C(r_\varphi) \,\forall t \in I_\varphi \,\forall G \in \mathcal{B}_\varphi : \Pi(t, \varphi + \psi, G) = 0 \iff t = \tau_\varphi(\varphi + \psi, G).$$

(6.7.7) *Claim:* The numbers $r_\varphi, \theta_\varphi, \beta_\varphi$ can be chosen uniformly for all $\varphi \in D$.

Proof: For $\varphi \in D$, set $D_\varphi := D_1 \Pi(\tau_0(\varphi), \varphi, F)$. An inspection of the proof of the Implicit Function Theorem ([Steinlein]) shows that it suffices to find $\bar{r}, \bar{\theta}, \bar{\beta} > 0$ with $C(\bar{r}) \subset U_1$, with $\bar{\mathcal{B}} := \{G \in BC^1 Lip(\mathcal{U}, \mathbb{R}^n) \mid \|G - F\|_{C^1} < \bar{\beta}\} \subset \mathcal{B}_1$, and $\bar{\theta} \in (0, \theta_1)$ such that the following two conditions hold for all $\varphi \in D, \psi \in C(\bar{r})$ and $G \in \bar{\mathcal{B}}$:

(6.7.8) $$|D_\varphi^{-1} \Pi(\tau_0(\varphi), \varphi + \psi, G)| \leq \bar{\theta}/2,$$

(6.7.9) $\quad \forall t \in \tau_0(\varphi) + (-\bar{\theta}, \bar{\theta}) : |D_\varphi^{-1} D_1 \Pi(t, \varphi + \psi, G) - \mathrm{id}_\mathbb{R}| \leq 1/2.$

Recall that for $(t, \chi, G) \in \mathcal{N}$, one has $t > 1$. Now boundedness of D, U_1 and of $\mathcal{B}_1$, equation (G) (for $G \in \mathcal{B}_1$), and the Arzelà-Ascoli Theorem show that the set $\mathrm{clos}(\Phi(\mathcal{N})) \subset C$ is compact. From (6.7.4), we have $\mathrm{clos}(\Phi(\mathcal{N})) \subset \mathcal{V}$. It follows that Dh is uniformly continuous on $\Phi(\mathcal{N})$, and hence there exists $\bar{\delta} > 0$ such that
(6.7.10)
$$\forall \chi_1, \chi_2 \in \Phi(\mathcal{N}) : |\chi_1 - \chi_2| \leq \bar{\delta} \implies |Dh(\chi_1) - Dh(\chi_2)| \leq \frac{d_0}{4(\|F\|_{C^0} + 1)}.$$

Choose $\bar{\theta} \in (0, \theta_1), \bar{r}, \bar{\beta} > 0$ such that $C(\bar{r}) \subset U_1$, that the set $\bar{\mathcal{B}}$ corresponding to $\bar{\beta}$ satisfies $\bar{\mathcal{B}} \subset \mathcal{B}_1$, and that the following inequalities hold, with $m := d_0/[24\mathrm{lip}(F)\|Dh\|_\infty]$.

(6.7.11)
$$\bar{\theta} \leq \min\{m, \bar{\delta}/3\}/(\|F\|_{C^0} + 1),$$
$$\bar{r} \leq \min\{m, \bar{\delta}/3, \frac{d_0 \bar{\theta}}{4\mathrm{lip}(h)}\}/\exp(\mathrm{lip}(F)T),$$
$$\bar{\beta} \leq \min\{m, \bar{\delta}/3, \frac{d_0}{8\|Dh\|_\infty}, \frac{d_0 \bar{\theta}}{4\mathrm{lip}(h)}, 1\}/T \exp(\mathrm{lip}(F)T).$$

We check that conditions (6.7.8) and (6.7.9) hold. Let $\varphi \in D, \psi \in C(\bar{r})$ and $G \in \bar{\mathcal{B}}$, and $t \in \tau_0(\varphi) + (-\bar{\theta}, \bar{\theta})$. Note first that, since $\|G\|_{C^0} \leq \|F\|_{C^0} + \bar{\beta} \leq \|F\|_{C^0} + 1$, we have the estimates
(6.7.12)
$$\Delta(\varphi + \psi, \varphi, G, F, t, \tau_0(\varphi)) \leq (\bar{r} + T\bar{\beta}) \exp(\mathrm{lip}(F)T) + (\|F\|_{C^0} + 1)\bar{\theta}$$
$$\leq 2m + m = d_0/[8\mathrm{lip}(F)\|Dh\|_\infty],$$

(6.7.13) $$\Delta(\varphi+\psi,\varphi,G,F,t,\tau_0(\varphi)) \leq \bar{\delta}/3 + \bar{\delta}/3 + \bar{\delta}/3 = \bar{\delta},$$

(6.7.14) $$\Delta(\varphi+\psi,\varphi,G,F,\tau_0(\varphi),\tau_0(\varphi)) \leq (\bar{r}+T\bar{\beta})\exp(\mathrm{lip}(F)T)$$
$$\leq 2\frac{d_0\bar{\theta}}{4\mathrm{lip}(h)} = \frac{d_0\bar{\theta}}{2\mathrm{lip}(h)}.$$

From the definition of D_φ and from (6.7.5), we know $|D_\varphi^{-1}| \leq d_0^{-1}$. Hence, since $\Pi(\tau_0(\varphi),\varphi,F) = 0$, we infer from (6.7.3) and (6.7.14)

$$|D_\varphi^{-1}\Pi(\tau_0(\varphi),\varphi+\psi,G)| \leq d_0^{-1}|\Pi(\tau_0(\varphi),\varphi+\psi,G) - \Pi(\tau_0(\varphi),\varphi,F)|$$
$$\leq d_0^{-1}\mathrm{lip}(h)\Delta(\varphi+\psi,\varphi,G,F,\tau_0(\varphi),\tau_0(\varphi)) \leq \bar{\theta}/2,$$

so condition (6.7.8) holds. Further,

$$|D_1\Pi(t,\varphi+\psi,G) - D_1\Pi(\tau_0(\varphi),\varphi,F)|$$
$$= |Dh(y_t^{\varphi+\psi,G})D_1\Phi_G(t,\varphi+\psi) - Dh(y_{\tau_0(\varphi)}^{\varphi,F})D_1\Phi_F(\tau_0(\varphi),\varphi)|$$
$$\leq |Dh(y_t^{\varphi+\psi,G}) - Dh(y_{\tau_0(\varphi)}^{\varphi,F})| \cdot |D_1\Phi_G(t,\varphi+\psi)| +$$
$$+ |Dh(y_{\tau_0(\varphi)}^{\varphi,F})| \cdot |D_1\Phi_G(t,\varphi+\psi) - D_1\Phi_F(\tau_0(\varphi),\varphi)|.$$

Combining (6.7.3), (6.7.10) and (6.7.13), we see that the first summand can be estimated by

$$\frac{d_0}{4(\|F\|_{C^0}+1)}\|G\|_{C^0} \leq d_0/4.$$

The second factor in the second summand equals $|G(y_\cdot^{\varphi+\psi,G})_t - F(y_\cdot^{\varphi,F})_{\tau_0(\varphi)}|$ and can be estimated by

(6.7.15) $$|G(y_\cdot^{\varphi+\psi,G})_t - F(y_\cdot^{\varphi+\psi,G})_t| + |F(y_\cdot^{\varphi+\psi,G})_t - F(y_\cdot^{\varphi,F})_{\tau_0(\varphi)}|$$
$$\leq \|G-F\|_{C^0} + \mathrm{lip}(F)\Delta(\varphi+\psi,\varphi,G,F,t,\tau_0(\varphi)).$$

Hence, using (6.7.12) and (6.7.11), we can estimate the second summand above by

$$\|Dh\|_\infty(\bar{\beta} + \mathrm{lip}(F)\frac{d_0}{8\mathrm{lip}(F)\|Dh\|_\infty}) \leq d_0/4.$$

Together, we have proved

(6.7.16) $$|D_1\Pi(t,\varphi+\psi,G) - D_1\Pi(\tau_0(\varphi),\varphi,F)| \leq d_0/2.$$

It follows now from (6.7.5) that

$$|D_\varphi^{-1}D_1\Pi(t,\varphi+\psi,G) - \mathrm{id}_\mathbb{R}|$$
$$= |D_1\Pi(\tau_0(\varphi),\varphi,F)^{-1}||D_1\Pi(t,\varphi+\psi,G) - D_1\Pi(\tau_0(\varphi),\varphi,F)|$$
$$\leq d_0^{-1}d_0/2 = 1/2.$$

Hence condition (6.7.9) holds, and claim (6.7.7) is proved.

With $\bar{r}, \bar{\theta}, \bar{\beta}$ and $\bar{\mathcal{B}}$ as given by claim (6.7.7), we now have for every $\varphi \in D$ a BC^1 map
$$\tau_\varphi : (\varphi + C(\bar{r})) \times \bar{\mathcal{B}} \to \tau_0(\varphi) + (-\bar{\theta}, \bar{\theta})$$
with the property

(6.7.17) $\quad \forall \psi \in C(\bar{r}) \, \forall G \in \bar{\mathcal{B}} \, \forall t \in \tau_0(\varphi) + (-\bar{\theta}, \bar{\theta}):$
$$\Pi(t, \varphi + \psi, G) = 0 \iff t = \tau_\varphi(\varphi + \psi, G).$$

(We prepare for the definition of the map τ now.) For $\varphi \in D$, $\psi \in C(\bar{r})$, $G \in \bar{\mathcal{B}}$ and $t \in \tau_0(\varphi) + (-\bar{\theta}, \bar{\theta})$, we have from (6.7.5) and (6.7.16) that $|D_1\Pi(t, \varphi + \psi, G)| \geq d_0/2$, so

(6.7.18) $\qquad\qquad |D_1\Pi(t, \varphi + \psi, G)^{-1}| \leq 2/d_0.$

Using this estimate, the derivative formula from the Implicit Function Theorem, and claim (6.7.2), we get the following estimate for the derivative of τ_φ:

$$|D\tau_\varphi(\varphi + \psi, G)|$$
$$= |D_1\Pi(\tau_\varphi(\varphi + \psi, G), \varphi + \psi, G)^{-1} D_{2,3}\Pi(\tau_\varphi(\varphi + \psi, G), \varphi + \psi, G)|$$
$$\leq \frac{2}{d_0} \|Dh\|_\infty \|D_{2,3}\Phi\|_\infty.$$

Abbreviating the last number with L, we obtain
(6.7.19)
$$|\tau_\varphi(\varphi + \psi, G) - \tau_0(\varphi)| = |\tau_\varphi(\varphi + \psi, G) - \tau_\varphi(\varphi, F)| \leq L(|\psi| + \|F - G\|_{C^1}).$$

Set now
$$\tilde{r} := \min\{\bar{\theta}/6L, \bar{r}/2\}, \quad \beta := \min\{\bar{\beta}, \bar{\theta}/6L\},$$
and define $\mathcal{B} := \{G \in BC^1 Lip(\mathcal{U}, \mathbb{R}^n) \mid \|G - F\|_{C^1} < \beta\}$. For $\varphi \in D, \psi \in C(\tilde{r}), G \in \mathcal{B}$, we have

(6.7.20) $\qquad |\tau_\varphi(\varphi + \psi, G) - \tau_0(\varphi)| \leq L(\bar{\theta}/6L + \bar{\theta}/6L) = \bar{\theta}/3.$

Since τ_0 is uniformly continuous, there exists $\delta_0 > 0$ such that
$$\forall \varphi_1, \varphi_2 \in D : |\varphi_1 - \varphi_2| \leq \delta_0 \implies |\tau_0(\varphi_1) - \tau_0(\varphi_2)| \leq \bar{\theta}/3.$$

We set $r := \min\{\tilde{r}, \delta_0/4\}$, and (as in the assertion of the theorem)
$$W := D + C(r).$$

(6.7.21) *Claim:* If $\varphi_i \in D, \psi_i \in C(2r)$ $(i = 1, 2)$, $G \in \mathcal{B}$, $\chi \in C$ and $\varphi_1 + \psi_1 = \chi = \varphi_2 + \psi_2$ then
$$\tau_{\varphi_1}(\chi, G) = \tau_{\varphi_2}(\chi, G).$$

Proof: We have
$$|\varphi_1 - \varphi_2| \leq |\varphi_1 + \psi_1 - (\varphi_2 + \psi_2)| + |\psi_1| + |\psi_2| \leq 0 + 4r \leq \delta_0,$$

so $|\tau_0(\varphi_1) - \tau_0(\varphi_2)| \leq \bar{\theta}/3$, and we obtain, using (6.7.20), that

$$|\tau_{\varphi_1}(\varphi_1 + \psi_1, G) - \tau_0(\varphi_2)| \leq |\tau_{\varphi_1}(\varphi_1 + \psi_1, G) - \tau_0(\varphi_1)| + |\tau_0(\varphi_1) - \tau_0(\varphi_2)|$$
$$\leq \bar{\theta}/3 + \bar{\theta}/3.$$

Thus we have

(6.7.22) $\qquad \tau_{\varphi_1}(\varphi_1 + \psi_1, G) \in \tau_0(\varphi_2) + [-2\bar{\theta}/3, 2\bar{\theta}/3],$

and, of course, $\varphi_1 + \psi_1 = \varphi_2 + \psi_2 \in \varphi_2 + C(2r) \subset \varphi_2 + C(2\tilde{r}) \subset \varphi_2 + C(\bar{r})$, and $G \in \mathcal{B} \subset \bar{\mathcal{B}}$. It follows from (6.7.17) that $\tau_{\varphi_1}(\varphi_1 + \psi_1, G) = \tau_{\varphi_2}(\varphi_2 + \psi_2, G)$. Claim (6.7.21) is proved.

In view of (6.7.21), we can now define $\tau : W \times \mathcal{B} \to \mathbb{R}$ by

$$\tau(\chi, G) := \tau_\varphi(\varphi + \psi, G) \text{ if } \chi = \varphi + \psi \text{ with } \varphi \in D, \psi \in C(r).$$

We have $\tau \in BC^1(W \times \mathcal{B}, \mathbb{R})$ (recall the passage before (6.7.19)). Set

$$\theta := \bar{\theta}/3.$$

Proof of assertion 1): For $\varphi \in D$ we have $\tau(\varphi, F) = \tau_\varphi(\varphi + 0, F) = \tau_0(\varphi)$ (compare (6.7.17) and assumption (iii)).

For $\chi = \varphi + \psi$, with $\varphi \in D, \psi \in C(r)$, and for $G \in B$, we have $\tau(\chi, G) \in \tau_0(\varphi) + (-\bar{\theta}, \bar{\theta})$, so (compare assumption (i))

$$\tau(\chi, G) > \tau_0(\varphi) - \bar{\theta} > \tau_0(\varphi) - \theta_0 > 1.$$

Further, using (6.7.20) and (6.7.1), we see that

$$\tau(\chi, G) + \theta \leq \tau_0(\varphi) + \bar{\theta}/3 + \theta = \tau_0(\varphi) + 2\bar{\theta}/3 < \tau_0(\varphi) + \theta_0 < t^+(\varphi + \psi, G).$$

Hence $\Phi_G(\cdot, \chi)$ is defined on $[0, \tau(\chi, G) + \theta]$, and (6.7.17) together with the definition of Π shows

$$0 = \Pi(\tau(\chi, G), \chi, G) = h[\Phi_G(\tau(\chi, G), \chi)].$$

Assume now that $t \in \tau(\chi, G) + (-\theta, \theta)$ and $h(\Phi_G(t, \chi)) = 0$. Then, from (6.7.20),
$$t \in \tau_0(\varphi) + (-\bar{\theta}/3, \bar{\theta}/3) + (-\theta, \theta) = \tau_0(\varphi) + (-2\bar{\theta}/3, 2\bar{\theta}/3).$$
It follows from (6.7.17) and the definition of τ that
$$t = \tau_\varphi(\varphi + \psi, G) = \tau(\chi, G).$$
Assertion 1) is proved; we turn to the proof of assertion 2). The chain rule and claim (6.7.2) imply that the maps P_G are BC^1. Set
$$\mathcal{O} := \bigcup_{\varphi \in D} (\tau_0(\varphi) + (-\bar{\theta}, \bar{\theta})) \times (\varphi + C(r)).$$

$\mathcal{O}$ is an open subset of $(1, \infty) \times C$, and we know from (6.7.2) that for every $G \in \mathcal{B}$ the map Φ_G is BC^1 on $\mathcal{O}$. For $G \in \mathcal{B}$, we have $\mathcal{O} \times \{G\} \subset \mathcal{M}$, and estimate (6.7.3) shows that Φ_G is uniformly continuous on $\mathcal{O}$.

Let now $\tilde{G} \in \mathcal{B}$ be given, with $D\tilde{G}$ uniformly continuous.
(6.7.23) *Claim:* $\Phi_G \to \Phi_{\tilde{G}}$ uniformly on $\mathcal{O}$, and $\tau(\cdot, G) \to \tau(\cdot, \tilde{G})$ uniformly on W as $\|G - \tilde{G}\|_{C^1} \to 0$.

Proof: It follows from (6.7.3) that for $(t, \chi) \in \mathcal{O}$ and $G \in \mathcal{B}$,
$$|\Phi_G(t, \chi) - \Phi_{\tilde{G}}(t, \chi)| = |y_t^{\chi, G} - y_t^{\chi, \tilde{G}}| \leq T \|G - \tilde{G}\|_{C^0} \exp(\mathrm{lip}(\tilde{G})T),$$
from which the first assertion is obvious. Recall that τ is BC^1; for $\chi \in W$ one has $|\tau(\chi, G) - \tau(\chi, \tilde{G})| \leq \|D_2 \tau\|_\infty \|G - \tilde{G}\|_{C^1}$. The second assertion follows.

(6.7.24) *Claim:* For $i = 1, 2$, $D_i \Phi_G \to D_i \Phi_{\tilde{G}}$ uniformly on $\mathcal{O}$ as $\|G - \tilde{G}\|_{C^1} \to 0$, and $D_i \Phi_{\tilde{G}}$ is uniformly continuous.

Proof: For $(t_i, \chi_i) \in \mathcal{O}$ $(i = 1, 2)$ and $G \in \mathcal{B}$, an estimate similar to (6.7.15) shows that
$$|D_1 \Phi_G(t_1, \chi_1) - D_1 \Phi_{\tilde{G}}(t_2, \chi_2)| \leq \|G - \tilde{G}\|_{C^0} + \mathrm{lip}(\tilde{G}) \Delta(\chi_1, \chi_2, G, \tilde{G}, t_1, t_2),$$
which yields the assertion for $i = 1$.

Recall the notation from the proof of claim (6.7.2), step 2. Let $G \in \mathcal{B}$ with $\|DG\|_\infty \leq \|D\tilde{G}\|_\infty + 1$, and $\eta \in C$, and assume first that $t_1 \leq t_2$. We obtain, using Proposition 6.2,

$|D_2 \Phi_G(t_1, \chi_1)\eta - D_2 \Phi_{\tilde{G}}(t_2, \chi_2)\eta|$
$\leq |w_{t_1}^{\chi_1, \eta, G} - w_{t_1}^{\chi_2, \eta, \tilde{G}}| + |w_{t_1}^{\chi_2, \eta, \tilde{G}} - w_{t_2}^{\chi_2, \eta, \tilde{G}}|$
$\leq |\eta| \cdot \|[DG(y^{\chi_1, G}) - D\tilde{G}(y^{\chi_2, \tilde{G}})]_{|[0, t_1]}\|_1 \exp(T\|D\tilde{G}\|_\infty) \exp(T\|DG\|_\infty) +$
$\quad + |t_2 - t_1| \cdot \|D\tilde{G}\|_\infty \max_{t \in [0, t_2]} |w_t^{\chi_2, \eta, \tilde{G}}|$
$\leq |\eta| \cdot \|[DG(y^{\chi_1, G}) - D\tilde{G}(y^{\chi_2, \tilde{G}})]_{|[0, t_1]}\|_1 \exp(T(2\|D\tilde{G}\|_\infty + 1)) +$
$\quad + |t_2 - t_1| \|D\tilde{G}\|_\infty |\eta| \exp(\|D\tilde{G}\|_\infty T).$

Together with an analogous estimate in case $t_1 > t_2$, we get in both cases
(6.7.25)
$$|D_2\Phi_G(t_1,\chi_1)\eta - D_2\Phi_{\tilde{G}}(t_2,\chi_2)\eta| \leq$$
$$\leq |\eta| \cdot \|[DG(y^{\chi_1,G}_{\cdot}) - D\tilde{G}(y^{\chi_2,\tilde{G}}_{\cdot})]|_{[0,\min\{t_1,t_2\}]}\|_1 \exp(T(2\|D\tilde{G}\|_\infty + 1)) +$$
$$|\eta| \cdot |t_2 - t_1|(\|D\tilde{G}\|_\infty + 1)\exp[(\|D\tilde{G}\|_\infty + 1)T].$$

Now

$$\|[DG(y^{\chi_1,G}_{\cdot}) - D\tilde{G}(y^{\chi_2,\tilde{G}}_{\cdot})]|_{0,\min\{t_1,t_2\}}\|_1$$
$$\leq \|DG - D\tilde{G}\|_{C^0} \min\{t_1,t_2\} +$$
$$+ \|[D\tilde{G}(y^{\chi_1,G}_{\cdot}) - D\tilde{G}(y^{\chi_2,\tilde{G}}_{\cdot})]|_{0,\min\{t_1,t_2\}}\|_1$$
$$\leq T\|DG - D\tilde{G}\|_{C^0} +$$
$$+ T\|[D\tilde{G}(y^{\chi_1,G}_{\cdot}) - D\tilde{G}(y^{\chi_2,\tilde{G}}_{\cdot})]|_{0,\min\{t_1,t_2\}}\|_{C^0([0,\min\{t_1,t_2\}],L_c(C,\mathbb{R}))}.$$

Uniform continuity of $D\tilde{G}$, the fact that $t_1, t_2 \leq T$, and estimate (6.3.1) show that the last term goes to zero as $\|G - \tilde{G}\|_{C^1} \to 0$ and $|\chi_1 - \chi_2| \to 0$. The assertion for $i = 2$ now follows from (6.7.25).

(6.7.26) *Claim:* $D\tau(\cdot, G) \to D\tau(\cdot, \tilde{G})$ uniformly on W as $\|G - \tilde{G}\|_{C^1} \to 0$.
Proof: For $\chi = \varphi + \psi \in W$ ($\varphi \in D, \psi \in C(r)$) and $G \in \mathcal{B}$, one has

(6.7.27)
$$D_1\tau(\chi, G) =$$
$$= -D_1\Pi(\tau(\chi,G),\chi,G)^{-1} \circ D_2\Pi(\tau(\chi,G),\chi,G)$$
$$= -\{Dh[\Phi_G(\tau(\chi,G),\chi)]D_1\Phi_G(\tau(\chi,G),\chi)\}^{-1} \circ$$
$$\circ Dh[\Phi_G(\tau(\chi,G),\chi)] \circ D_2\Phi_G(\tau(\chi,G),\chi).$$

Note that from (6.7.18) we have

(6.7.28)
$$\sup_{\chi \in W, G \in \mathcal{B}} |\{...\}^{-1}| \leq 2/d_0.$$

Note further that $\Phi_G(\tau(\chi,G),\chi) \in \Phi(\mathcal{N})$, and recall that Dh is uniformly continuous on the latter set. Combining these properties with claims (6.7.23) and (6.7.24), and with boundedness of Dh and of $D_i\Phi_G$ ($i = 1, 2$) (uniformly for all $G \in \mathcal{B}$), one obtains the assertion.

(6.7.29) *Claim:* The maps $\tau(\cdot, \tilde{G})$ and $D\tau(\cdot, \tilde{G})$ are uniformly continuous on W.

Proof: For $\varphi_1, \varphi_2 \in D$, $\psi_1, \psi_2 \in C(r)$ with $|\varphi_1 + \psi_1 - (\varphi_2 + \psi_2)| < r$, one has $\psi := \varphi_2 + \psi_2 - \varphi_1 \in C(2r)$, so (6.7.21) shows that

$$\tau_{\varphi_2}(\varphi_2 + \psi_2, \tilde{G}) = \tau_{\varphi_2}(\varphi_1 + \psi, \tilde{G}) = \tau_{\varphi_1}(\varphi_1 + \psi, \tilde{G}) = \tau_{\varphi_1}(\varphi_2 + \psi_2, \tilde{G}).$$

From the definitions of L and τ, we thus obtain, using the Intermediate Value Theorem,

$$\begin{aligned}|\tau(\varphi_1 + \psi_1, \tilde{G}) - \tau(\varphi_2 + \psi_2, \tilde{G})| &= |\tau_{\varphi_1}(\varphi_1 + \psi_1, \tilde{G}) - \tau_{\varphi_2}(\varphi_2 + \psi_2, \tilde{G})| \\ &= |\tau_{\varphi_1}(\varphi_1 + \psi_1, \tilde{G}) - \tau_{\varphi_1}(\varphi_2 + \psi_2, \tilde{G})| \\ &\leq L|\varphi_1 + \psi_1 - (\varphi_2 + \psi_2)|.\end{aligned}$$

In particular, $\tau(\cdot, \tilde{G})$ is uniformly continuous on W.

Proof of the second assertion: We know that $\Phi_{\tilde{G}}$ is uniformly continuous on $\mathcal{O}$ (see the passage preceding claim (6.7.23)). Consider now formula (6.7.27) for the case $G = \tilde{G}$. Property (6.7.28), boundedness of $D_i \Phi_{\tilde{G}}$ ($i = 1, 2$) and of Dh, and uniform continuity of Dh on $\Phi(\mathcal{N})$, and uniform continuity of $\Phi_{\tilde{G}}$, of $\tau(\cdot, \tilde{G})$, and of $D_1 \Phi_{\tilde{G}}, D_2 \Phi_{\tilde{G}}$ (see (6.7.24)), all combined, yield the second assertion of the claim.

We can now complete the proof of assertion 2) of the theorem. For $G \in \mathcal{B}$, define $f_{1,G} : W \to \mathcal{O} \subset \mathbb{R} \times C$,

$$f_{1,G}(\varphi + \psi) := (\tau(\varphi + \psi, G), \varphi + \psi),$$

and define

$$f_{2,G} := \Phi_G|_{\mathcal{O}} : \mathcal{O} \to C.$$

(Note that $P_G = f_{2,G} \circ f_{1,G}$.) These maps are BC^1; we want to apply Proposition 6.6 with $n := 2$, with $\Lambda := \mathcal{B}$ and $\lambda_0 := \tilde{G}$. Condition a) of Proposition 6.6 holds in view of claim (6.7.23), of the first assertion in (6.7.24) and claim (6.7.26). From uniform continuity of $\Phi_{\tilde{G}}$ on $\mathcal{O}$, from the second assertion in (6.7.24), and from (6.7.29), we see that condition b) of Proposition 6.6 also holds, even for $n = 1$. The conclusion obtained from Proposition 6.6 is that $\|P_G - \tilde{P}_G\|_{C^1} \to 0$ as $\|G - \tilde{G}\|_{C^1} \to 0$, and that $DP_{\tilde{G}}$ is uniformly continuous. □

The following lemma is a specialization of much more general results which have been known, at least, since the work of Wright ([Wright 1], [Wright 2]). The generality of such theorems makes it somewhat inconvenient to interpret their content in the extremely simple case that is important here.

For the proof, we indicate corresponding passages in the literature, together with simple additional considerations that specialize the more general results. A self-contained, completely detailed proof for the equation $\dot{x}(t) = \alpha x(t - 1)$ is available upon request, but was not included here, regarding the literature on this issue.

Recall the complexification of C, namely $C_{\mathbb{C}} = \{\varphi + i\psi \mid \varphi, \psi \in C\}$.

6.8. Lemma (Characteristic values, eigenspaces, series expansions). *Let $\alpha > 0$.*

a) The zeroes of the characteristic function $\mathbb{C} \ni z \mapsto z - \alpha e^{-z}$ associated to equation (α) $\dot{x}(t) = \alpha x(t-1)$ are simple. The zero set Z has the form

$$Z = \{\lambda\} \cup \{\mu_k \mid k = 1, 2, ...\} \cup \{\bar{\mu}_k \mid k = 1, 2, ...\},$$

where $\lambda \in \mathbb{R}$, $\lambda > 0$ and $\mu_k = \rho_k + i\omega_k$, $\omega_k \neq 0$. With

$$\chi(\omega) := \frac{\omega}{\sin \omega} \exp(-\omega \cot \omega)$$

for $\omega \in \{z \in \mathbb{C} \mid \sin z \neq 0\}$, one has for $k = 1, 2, ...$

$$\chi(\omega_k) = -\alpha < 0, \quad \omega_k \in ((2k-1)\pi, 2k\pi),$$

and $\chi' < 0$ on this interval. Further, for $k \in \mathbb{N}$,

$$\rho_k = -\omega_k \cot \omega_k, \quad \rho_{k+1} < \rho_k,$$

and $\alpha \in (0, 3\pi/2) \Longrightarrow \rho_k < 0$ for all $k \in \mathbb{N}$.

b) For $\mu \in Z$ and $\psi \in C_\mathbb{C}$, define

$$\mathrm{pr}_\mu \psi := \frac{1}{1+\mu} [\psi(0) + \mu \int_{-1}^{0} e^{-\mu s} \psi(s) ds].$$

The subspaces $E_\mu := \mathbb{C} \cdot \exp(\mu \cdot)_{|[-1, 0]}$ of $C_\mathbb{C}$ are invariant under the semigroup $T_\mathbb{C}^\alpha : \mathbb{R}_0^+ \to L_c(C_\mathbb{C}, C_\mathbb{C})$ induced by equation (α). For $\psi \in E_\mu$ and $t \geq 0$, one has $T_\mathbb{C}^\alpha(t)\psi = \exp(\mu t)\psi$. The spectral projections onto these spaces are given by

$$\pi_\mu \psi = \mathrm{pr}_\mu \psi \cdot \exp(\mu \cdot)_{|[-1, 0]} \quad (\psi \in C_\mathbb{C}, \mu \in Z).$$

In particular, $\pi_{\mu_1} \pi_{\mu_2} = 0$ for $\mu_1, \mu_2 \in Z$, $\mu_1 \neq \mu_2$. For $\mu \in Z$, the space $\ker \pi_\mu$ is also invariant under the semigroup $T_\mathbb{C}^\alpha$.

c) If $k \in \mathbb{N}$ and $\gamma > \rho_{k+1}$ then, with $S_\mathbb{C} := \ker \pi_\lambda \cap \bigcap_{j=1}^{k} \ker \pi_{\mu_j} \cap \bigcap_{j=1}^{k} \ker \pi_{\bar{\mu}_j}$,

one has $C_\mathbb{C} = S_\mathbb{C} \oplus \bigoplus_{j=1}^{k} (E_{\mu_j} \oplus E_{\bar{\mu}_j}) \oplus E_\lambda$, and there exists $K > 0$ such that for all $\psi \in S_\mathbb{C}$, $t \geq 0$:

$$|T_\mathbb{C}^\alpha(t)\psi| \leq K \exp(\gamma t).$$

The subspaces $\mathrm{Re}(S_\mathbb{C})$, $\bigoplus_{j=1}^{k} \mathrm{Re}(E_{\mu_j})$, $\mathrm{Re}(E_\lambda)$ of the above spaces form a direct sum decomposition of C which is invariant under the semigroup T^α induced on C, with an analogous estimate on $\mathrm{Re}(S_\mathbb{C})$.

d) *For $\psi \in C_{\mathbb{C}}$, the solution $y^\psi : [-1, \infty) \to \mathbb{C}$ of equation (α) with $y_0^\psi = \psi$ satisfies*

$$\forall t > 0 : y^\psi(t) = \sum_{\mu \in Z} (\mathrm{pr}_\mu \psi) \exp(\mu t).$$

Outline of the proof. The proof of a) is a slight modification of the proof of Theorem 5 in [Wright 2], p. 72, taking into account that $\alpha > 0$ here.

Ad b): Z is the set of eigenvalues of the infinitesimal generator $A_{\mathbb{C}}$ of $T_{\mathbb{C}}^\alpha$. The formula for pr_μ can be obtained from the general Cauchy integral formula ([Taylor, Lay], p. 310) for spectral projections and from explicit computation of the resolvent $(z - A_{\mathbb{C}})^{-1}$ for $z \in \mathbb{C} \setminus Z$. (The formula is a special case of, e.g., formula (3.3) from [Diekmann et al.], p. 106.) Spectral projections π_1, π_2 corresponding to disjoint, closed subsets of the spectrum generally satisfy $\pi_1 \pi_2 = 0$. Invariance of E_μ ($\mu \in Z$) under $T_{\mathbb{C}}^\alpha$ is clear, since $t \mapsto \exp(\mu t)$ defines a solution of equation (α) for all $\mu \in Z$.

Ad c): The decomposition follows from general spectral theory (e.g., [Taylor, Lay], p. 320-321, 309-312, 287- 297), and the estimate is a special case of, e.g., Theorem 6.1 from [Hale, Verduyn Lunel], p. 214.

Ad d): The series expansion is a special case of Theorem 4.1, c) in [Bellman, Cooke], p. 107. The latter result, applied to $u(t) = y^\psi(t - 1)$ ($t \geq 0$), yields the formula

$$u(t) = \sum_{\mu \in Z} \mathrm{Res}_\mu [s \mapsto \frac{e^{ts} p(s)}{s - \alpha e^{-s}}] \text{ for } t > 1,$$

where

$$p(s) = u(1)e^{-s} + s \int_0^1 u(\tau) e^{-s\tau} d\tau$$

$$= y^\psi(0) e^{-s} + s \int_{-1}^0 y^\psi(\tau) e^{-s(\tau+1)} d\tau$$

$$= e^{-s} [\psi(0) + s \int_{-1}^0 \psi(\tau) e^{-s\tau} d\tau].$$

Hence, for $t > 0$,

$$y^\psi(t) = u(t+1) = \sum_{\mu \in Z} \mathrm{Res}_\mu [s \mapsto \frac{e^{ts}(\psi(0) + s \int_{-1}^0 \psi(\tau) e^{-s\tau} d\tau)}{s - \alpha e^{-s}}].$$

Since all $\mu \in Z$ are simple zeroes of the denominator, the expression above equals

$$\sum_{\mu \in Z} [\psi(0) + \mu \int_{-1}^0 \psi(\tau) e^{-\mu \tau} d\tau] \cdot e^{\mu t} \cdot \mathrm{Res}_\mu [s \mapsto \frac{1}{s - \alpha e^{-s}}].$$

Now, for $\mu \in Z$, $\mu = \alpha e^{-\mu}$ implies that

$$\operatorname{Res}_\mu \frac{1}{s - \alpha e^{-s}} = \operatorname{Res}_\mu \left(\frac{1}{s - \mu} \frac{1}{1 - \alpha \frac{e^{-s} - e^{-\mu}}{s - \mu}} \right) = \frac{1}{1 - \alpha \frac{d}{ds} e^{-s}}\bigg|_{s = \mu}$$

$$= \frac{1}{1 + \alpha e^{-\mu}} = \frac{1}{1 + \mu}.$$

It follows that, for $t > 0$,

$$y^\psi(t) = \sum_{\mu \in Z} \frac{1}{1 + \mu} [\psi(0) + \mu \int_{-1}^{0} \psi(\tau) e^{-\mu \tau} d\tau] \cdot e^{\mu t}.$$

In view of the definition of pr_μ, this is the asserted expansion. □

In the last statement of this section, we calculate some estimates for values of analytic functions, basically using Taylor expansion. The estimates, although hidden here at the end of the appendix, are crucial for the proof of existence of a heteroclinic solution with the required properties in Section 5.

6.9. Proposition (Numerical estimates).
a) $1.09 \leq \log 3 \leq 1.1$
b) $\log 2 \leq 0.7$
c) $3.31 \leq 9/e \leq 3.3113$
d) For $\lambda = \lambda^* = 2 \log 3 - 1$ and $\alpha = \alpha^* = \lambda^* e^{\lambda^*}$, one has

$$3.9058 \leq \alpha \leq 3.9744,$$
$$\alpha \lambda \leq 4.77.$$

e) Let $\mu_1 = \rho_1 + i\omega_1$ be the solution of the characteristic equation $z = \alpha^* e^{-z}$ with largest negative real part, and with $\omega_1 > 0$. Then

$$3\pi/2 > \omega_1 > 4.67, \text{ and } -0.21 \leq \rho_1 < 0.$$

Proof. Ad a):

$$e^{1.1} \geq 1 + 1.1 + \frac{1.21}{2} + \frac{1.331}{6} + \frac{1.461}{24} + \frac{1.61}{120}$$
$$\geq 2.1 + 0.605 + 0.2218 + 0.061 + 0.013$$
$$= 3.0008 > 3,$$

so $\log 3 < 1.1$. Further, $e^{0.09} = 1 + 0.09 + \frac{0.09^2}{2} + e^\xi \frac{0.09^3}{6}$, with some $\xi \in [0, 0.09]$, and $e^\xi \leq e \leq 3$. It follows that

$$e^{0.09} \leq 1.09 + \frac{0.0081}{2} + \frac{0.09^3}{2} = 1.09 + 0.0405 + \frac{0.000729}{2} = 1.1008645$$
$$\leq 1.1009,$$

and $e^{1.09} \leq 2.7183 \cdot 1.1009 = 2.99257647 < 3$, so $\log 3 > 1.09$.

Ad b): $e^{0.7} > 1 + 0.7 + \dfrac{0.49}{2} + \dfrac{0.343}{6} \geq 1.945 + 0.057 > 2$.

Ad c):
$$\frac{9}{e} \leq \frac{9}{2.718} = \frac{10}{3.02} \leq 3.3113.$$
$$\frac{9}{e} \geq \frac{9}{2.7183} \geq \frac{10}{3.02033\ldots} \geq \frac{10}{3.0204} \geq 3.31.$$

Ad d): From a) and c) we see that
$$\alpha = \lambda e^\lambda = \lambda \frac{9}{e} \leq (2 \cdot 1.1 - 1) \cdot 3.3113 \leq 1.2 \cdot 3.312 = 3.9744,$$
$$\alpha \geq 1.18 \cdot 3.31 = 3.9058.$$

Finally, $\alpha \lambda \leq 3.9744 \cdot 1.2 = 4.76928 \leq 4.77$.

Ad e): From Lemma 6.8, a). we know that with $\chi(\omega) := \frac{\omega}{\sin \omega} e^{-\omega \cot \omega}$ we have $\chi(\omega_1) = -\alpha$, and $\chi' < 0$ on $(\pi, 2\pi)$. Note that
$$\lim_{\omega \to \pi+} \chi(\omega) = 0, \; \chi(\omega) \to -\infty \; (\omega \to 2\pi).$$

Claim: $|\chi(4.67)| < \alpha$.

Proof: Set $\delta := 3\pi/2 - 4.67$. Then
$$\delta \geq \frac{3.1415 \cdot 3}{2} - 4.67 = \frac{9.4245}{2} - 4.67 = 4.71225 - 4.67 = 0.04225,$$
$$\delta \leq \frac{3.1416 \cdot 3}{2} - 4.67 = \frac{9.4248}{2} - 4.67 = 4.7124 - 4.67 = 0.0424.$$

We have
$$0.0424 \geq \delta \geq \sin(\delta) \geq \delta - \frac{\delta^3}{6} \geq 0.04225 - \frac{0.05^3}{6}$$
$$= 0.04225 - \frac{0.000125}{6} \geq 0.04225 - 0.000021$$
$$\geq 0.04222.$$

Hence,
$$|\cos(4.67)| = |-\cos(4.67 - \pi)| = |-\sin(3\pi/2 - 4.67)| = |-\sin \delta| \in [0.04222, \delta].$$

Using $\sqrt{1+x} \leq 1 + x/2$ for $x \geq 0$, and $\delta^2 \leq 0.01$, we get
$$\frac{1}{|\sin 4.67|} \leq \frac{1}{\sqrt{1-\delta^2}} = \sqrt{1 + \frac{\delta^2}{1-\delta^2}} \leq 1 + \frac{\delta^2}{2(1-\delta^2)} \leq 1 + \frac{\delta^2}{1.98}$$
$$\leq 1 + \frac{0.043^2}{1.98} \leq 1 + \frac{1849}{198} 10^{-4} \leq 1.001.$$

It follows that $\dfrac{4.67}{|\sin 4.67|} \leq 4.67 \cdot 1.001 = 4.67467 \leq 4.675$. Now,

$$4.67|\cot 4.67| \geq 4.67|\cos 4.67| \geq 4.67 \cdot 0.042 \geq 0.196.$$

Since $\pi < 4.67 < 3\pi/2$, we have $\sin 4.67 < 0$ and $\cos 4.67 < 0$, so

$$|e^{-4.67 \cot 4.67}| = e^{-4.67|\cot 4.67|} \leq e^{-0.196}$$
$$\leq \frac{1}{1 + 0.196 + \frac{0.196^2}{2}} \leq \frac{1}{1.2} \leq 0.834.$$

Together, we obtain

$$|\chi(4.67)| = \frac{4.67}{|\sin 4.67|}|e^{-4.67 \cot 4.67}| \leq 4.675 \cdot 0.834 = 3.89895.$$

From assertion d), we now see that $|\chi(4.67)| < 3.9058 < \alpha$. The claim is proved, and it follows that
$$\omega_1 > 4.67.$$

Since $\chi(3\pi/2) = -3\pi/2 < -\alpha$, we have $\omega_1 < 3\pi/2$, and $\cot \omega_1 > 0$. Now $\cot(3\pi/2) = 0$ and $\cot' = -1/\sin^2 < 0$, so we have $\cot \omega_1 \leq \cot 4.67$. We see from Lemma 6.8, a) that $\rho_1 = -\omega_1 \cot(\omega_1) < 0$. Further (compare the upper estimate for δ), we have

$$|\rho_1| \leq \frac{3\pi}{2}|\cot \omega_1| \leq \frac{3\pi}{2}|\cot 4.67| \leq \frac{3\pi}{2} 1.001 \cdot \delta$$
$$\leq 4.7124 \cdot 1.001 \cdot 0.0424 \leq 0.19980576 \cdot 1.001 \leq 0.21.$$

□

REFERENCES

[An der Heiden, Walther] U. An der Heiden and H.-O. Walther, *Existence of Chaos in Control Systems with Delayed Feedback*, J. Differential Equations **47** (1983), 273–295.

[Bellman, Cooke] R. Bellman and K.L. Cooke, *Differential-Difference Equations*, Academic Press, 1963.

[Benedicks, Carleson 1] M. Benedicks and L. Carleson, *On iterations of $1 - ax^2$ on $[-1, 1]$*, Ann. of Math. **122** (1985), 1–25.

[Benedicks, Carleson 1] M. Benedicks and L. Carleson, *Dynamics of the Hénon map*, Ann. of Math. **133** (1990), 73–169.

[Birkhoff] G.D. Birkhoff, *Dynamical Systems*, American Mathematical Society, New York, 1927.

[Diekmann et al.] O. Diekmann, S.A. van Gils, S.M. Verdun-Lunel, and H.-O. Walther, *Delay Equations*, (Applied Mathematical Sciences 110) Springer–Verlag, New York, 1995.

[Dormayer 1] P. Dormayer, *Smooth symmetrybreaking bifurcation for functional differential equations*, Differential Integral Equations **5 No. 4** (1992), 831–854.

[Dormayer 2] P. Dormayer, *Floquet multipliers and Secondary Bifurcation of Periodic Solutions of Functional Differential Equations*, habilitation thesis, Mathematisches Institut der Universität Giessen, 1996.

[Dormayer, Lani-Wayda] P. Dormayer and B. Lani–Wayda, *Floquet multipliers and Secondary Bifurcations in Functional Differential Equations. Numerical and Analytical Results*, Z. Angew. Math. Phys. (ZAMP) **46** (1995), 823–858.

[Driver] R.D. Driver, *Ordinary and delay differential equations*, (Applied Mathematical Sciences 20) Springer Verlag, New York, 1977.

[Faria, Magalhães] T. Faria and L.T. Magalhães, *Realisation of ordinary differential equations by retarded functional–differential equations in neighbourhoods of equilibrium points*, Proc. Roy. Soc. Edinburgh Sect. A **125 no. 4** (1995), 759–776.

[Furumochi] T. Furumochi, *Existence of periodic solutions of one-dimensional differential–delay equations*, Tôhoku Math. J. **30** (1978), 13–35.

[Guckenheimer, Holmes] J. Guckenheimer and Ph. Holmes, *Nonlinear Oscillations, Dynamical Systems, and Bifurcations of Vector Fields, 2nd edition*, Springer Verlag, New York, 1986.

[Hadamard] J. Hadamard, *Les surfaces à courbures opposées et leur lines géodésiques*, J. de Math. Pures Appl. (5) **4** (1898), 27-74.

[Hale, Lin] J.K. Hale and X.B. Lin, *Examples of transverse homoclinic orbits in delay equations*, Nonlinear Anal. **10** (1986), 693–709.

[Hale, Sternberg] J.K. Hale and N. Sternberg, *Onset of Chaos in Differential Delay Equations*, J. Comput. Phys. **77 No. 1** (1988), 221–239.

[Hale, Verduyn Lunel] J.K. Hale and S.M. Verduyn Lunel, *Introduction to Functional Differential Equations*, (Applied Mathematical Sciences 99) Springer Verlag, New York, 1993.

[Holmes] P.J. Holmes, *A strange family of three–dimensional vector fields near a degenerate singularity*, J. Differential Equations **37** (1980), 382–404.

[Kaplan, Yorke]   J.L. Kaplan and J.A. Yorke, *Ordinary differential equations which yield periodic solutions of differential delay equations*, J. Math. Anal. Appl. **48** (1974), 317–324.

[Kirchgraber, Stoffer]   U. Kirchgraber and D. Stoffer, *Chaotic behavior in simple dynamical systems*, SIAM Rev. **32 No.3** (1990), 424-452.

[Lani-Wayda 1]   B. Lani-Wayda, *Quasi-random behavior in the neighborhood of a homoclinic orbit*, diploma thesis, Mathematisches Institut der Universität München, 1989.

[Lani-Wayda 2]   B. Lani-Wayda, *Hyperbolic Sets, Shadowing and Persistence for Noninvertible Mappings in Banach spaces*, Pitman Research Notes in Mathematics No. 334, Longman Group Ltd., Harlow, Essex, 1995.

[Lani-Wayda 3]   B. Lani-Wayda, *Persistence of Poincaré mappings in functional differential equations (with application to structural stability of complicated behavior)*, J. Dynam. Differential Equations **7 No. 1** (1995), 1–71.

[Lani-Wayda 4]   B. Lani-Wayda, *Erratic solutions of simple delay equations*, Trans. Amer. Math. Soc. **351** (1999), 901–945.

[Lani-Wayda, Walther 1]   B. Lani-Wayda and H.-O. Walther, *Chaotic motion generated by delayed negative feedback, Part I: A transversality criterion*, Differential Integral Equations **8 No. 6** (1995), 1407–1452.

[Lani-Wayda, Walther 2]   B. Lani-Wayda and H.-O. Walther, *Chaotic motion generated by delayed negative feedback, Part II: Construction of nonlinearities*, Math. Nachr. **180** (1996), 141–211.

[Lasota]   A. Lasota, *Ergodic problems in biology*, Asterisque **50** (1977), 239–250.

[Lasota, W.-Czyzewska]   A. Lasota and M. Wazewska–Czyzewska, *Matematyczne problemy dynamiki ukladu krwinek czerwonych*, Mat. Stos. **6** (1976), 23–40.

[Lazutkin]   V.A. Lazutkin, *Positive Entropy for the Standard Map I* (1994), Preprint 94-47, Université de Paris-Sud, Mathématiques Bâtiment 425, 91405 Orsay, France.

[Li, Yorke]   T.Y. Li and J.A. Yorke, *Period three implies chaos*, Amer. Math. Monthly **82** (1977), 985–99.

[Lin]   X.B. Lin, *Using Melnikov's method to solve Šilnikov's problems*, Proc. Roy. Soc. Edinburgh Sect. A **116** (1990), 295–325.

[Mackey, Glass]   M.C. Mackey and L. Glass, *Oscillation and chaos in physiological control systems*, Science **197** (1977), 287–285.

[McLaughlin, Shatah]   D.W. McLaughlin and J. Shatah, *Melnikov Analysis for Pde's*, (1996), in: P. Deift P., C.D. Levermore and C.E. Wayne (eds.), Dynamical Systems and Probabilistic Methods in Partial Differential Equations, Lectures in Applied Mathematics Vol. 31, American Mathematical Society, Providence, RI.

[Mallet-Paret]   J. Mallet–Paret, *Morse decompositions for differential delay equations*, J. Differential Equations **72** (1988), 270–315.

[Mallet-Paret, Sell]   J. Mallet–Paret and G. Sell, *The Poincaré–Bendixson theorem for monotone cyclic feedback systems with delay*, J. Differential Equations **125** (1996), 441–489.

[Mischaikow, Mrozek]   K. Mischaikow and M. Mrozek, *Isolating neighborhoods and chaos*, Japan J. Industr. Appl. Math. **12 No. 2** (1995), 205–236.

[Morse]   M. Morse, *A one-to-one representation of geodesics on a surface of negative curvature*, Amer. J. Math. **43** (1921), 33–51.

REFERENCES

[Morse, Hedlund]   M. Morse and G. Hedlund, *Symbolic Dynamics*, Amer. J. Math. **60** (1938), 815–866.

[Moser]   J. Moser, *Stable and Random Motions in Dynamical Systems*, Annals of Math. Studies No. 77, Princeton University Press, Princeton, New Jersey, 1973.

[Nussbaum]   R.D. Nussbaum, *Periodic solutions of some nonlinear autonomous functional differential equations*, Ann. Mat. Pura Appl. **101** (1974), 263–306.

[Palmer]   K.J. Palmer, *Exponential dichotomies, the shadowing lemma and transversal homoclinic points.* (1988), In: U. Kirchgraber and H.-O. Walther (eds.), Dynamics Reported, Vol. I, Teubner-Wiley, Stuttgart/Chichester.

[Peters]   H. Peters, *Globales Lösungsverhalten zeitverzögerter Differentialgleichungen am Beispiel von Modellfunktionen*, Doctoral Dissertation, Universität Bremen, 1980.

[Poincaré]   H. Poincaré, *Méthodes nouvelles de la Mécanique Céleste, Tome III*, Gauthier-Villars, Paris, 1899.

[Poláčik]   P. Poláčik, *Imbedding of any vector field in a scalar semilinear parabolic equation*, Proc. Amer. Math. Soc. **115 No. 4** (1992), 1001–1008.

[Rybakowski]   K.P. Rybakowski, *Realization of Arbitrary Vector Fields on Invariant Manifolds of Delay Equations*, J. Differential Equations **114** (1994), 222–231.

[Sandstede, Fiedler]   B. Sandstede and B. Fiedler, *Dynamics of periodically forced parabolic equations on the circle*, Ergodic Theory Dynam. Systems **12 No. 3** (1992), 559–571.

[Šil'nikov 1]   L.P. Šil'nikov, *A case of the existence of a denumerable set of periodic motions*, Sov. Math. Dokl. **6** (1965), 163–166.

[Šil'nikov 2]   L.P. Šil'nikov, *On a Poincaré–Birkhoff Problem*, Sb. Math. **3** (1967), 353–371.

[Smale]   S. Smale, *Differentiable Dynamical Systems*, Bull. Amer. Math. Soc. **73** (1967), 747–817.

[Steinlein]   H. Steinlein, *Nichtlineare Funktionalanalysis I*, (lecture script) Universität München, winter semester 1986/87.

[Steinlein, Walther]   H. Steinlein and H.-O. Walther, *Hyperbolic Sets, Transversal Homoclinic Trajectories, and Symbolic Dynamics for C^1-maps in Banach Spaces*, J. Dynam. Differential Equations **2** (1990), 325–365.

[Taylor, Lay]   A.E. Taylor and D.C. Lay, *Introduction to Functional Analysis*, 2nd edition, Wiley, New York, 1980.

[Walther 1]   H.-O. Walther, *Homoclinic solution and chaos in $\dot{x}(t) = f(x(t-1))$*, Nonlinear Anal. **5** (1981), 775–788.

[Walther 2]   H.-O. Walther, *Bifurcation from periodic solutions in in functional differential equations*, Math. Z. **182** (1983), 269-289.

[Walther 3]   H.-O. Walther, *Hyperbolic periodic solutions, heteroclinic connections and transversal homoclinic points in autonomous differential delay equations*, Mem. Amer. Math. Soc. **402** (1989), American Mathematical Society, Providence, RI.

[Walther 4]   H.-O. Walther, *Homoclinic and periodic solutions of scalar differential delay equations*, Dynamical Systems and Ergodic Theory, Banach Center Publications **23** (1989), 243–264.

[Wiggins]   S. Wiggins, *Global Bifurcations and Chaos*, Springer-Verlag, New York, 1988.

[Wischert et al.] W. Wischert, A. Wunderlin, A. Pelster, M. Olivier, J. Groslambert, *Delay–induced instabilities in nonlinear feedback systems*, Phys. Rev. E **49 No. 1** (1994), 203–219.

[Wright 1] E.M. Wright, *The linear difference-differential equation with constant coefficients*, Proc. Roy. Soc. Edinburgh Sect. A **62** (1949), 387–393.

[Wright 2] E.M. Wright, *A non-linear difference-differential equation*, J. Reine Angew. Math. **194** (1955), 66–87.

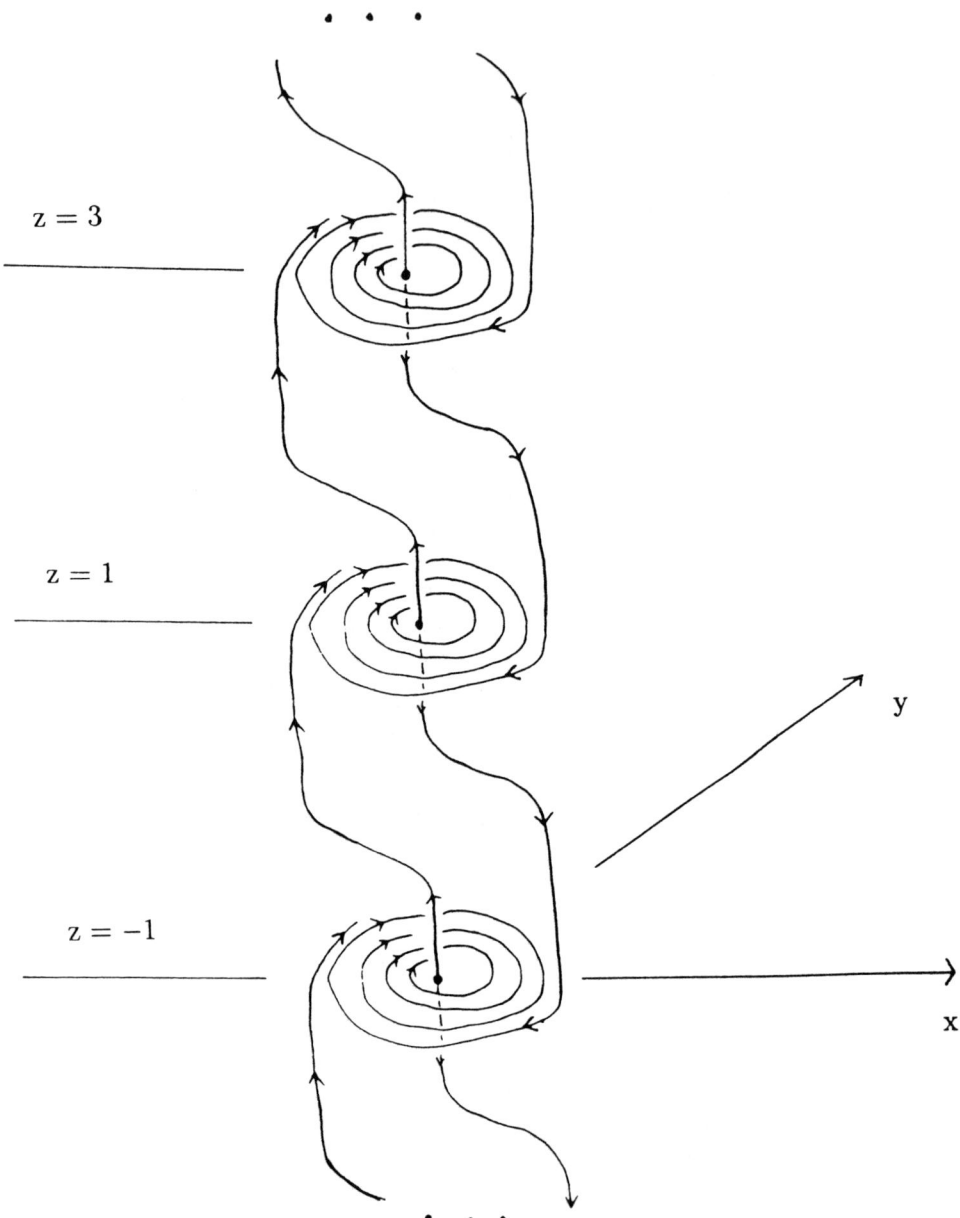

FIGURE 1. A sequence of connected spiral saddles

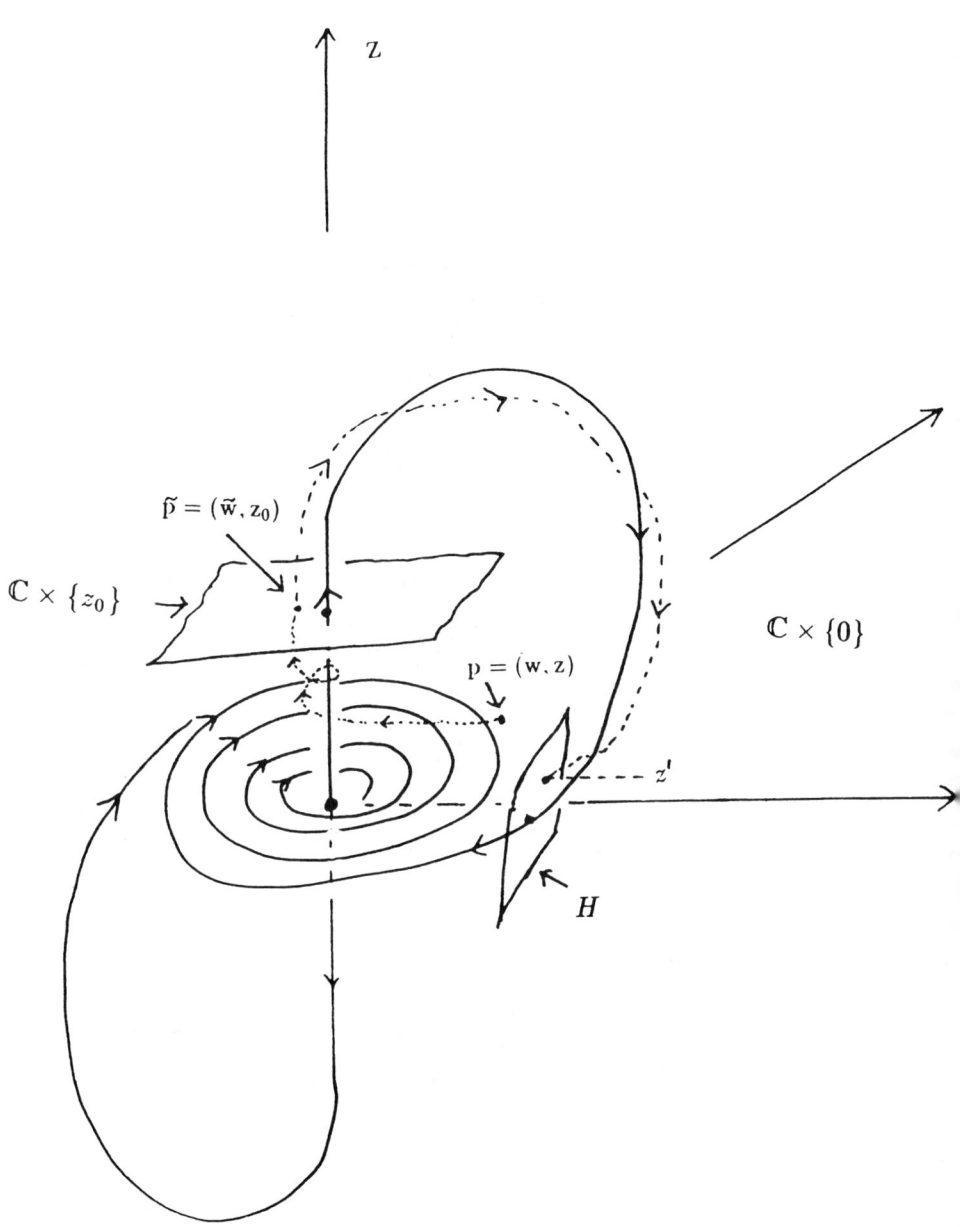

FIGURE 2. A spiral saddle with both branches of the unstable manifold homoclinic

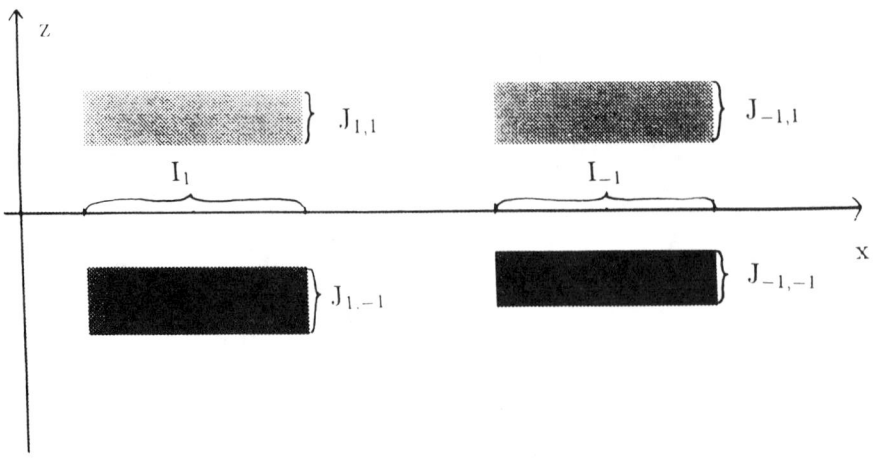

FIGURE 3. The set $\Delta = \bigcup_{i,j=-1,1} (I_i \times J_{i,j})$

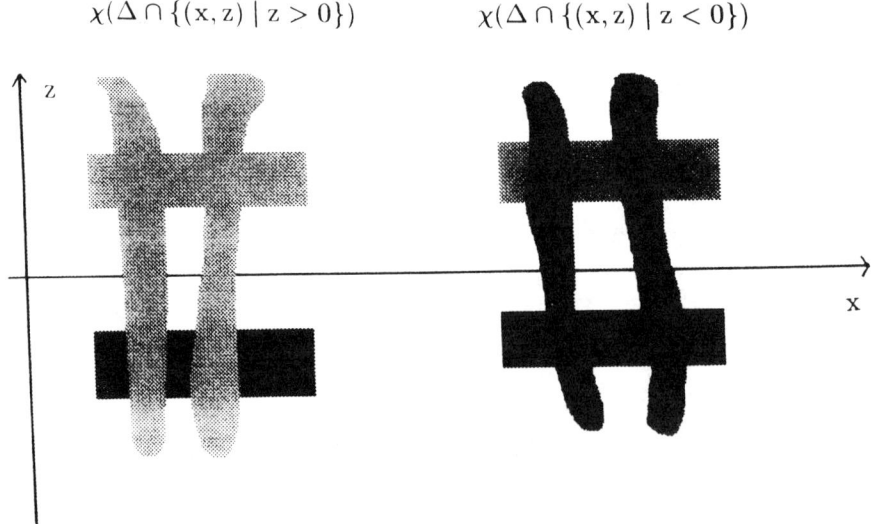

FIGURE 4. The set Δ and its image under the map χ

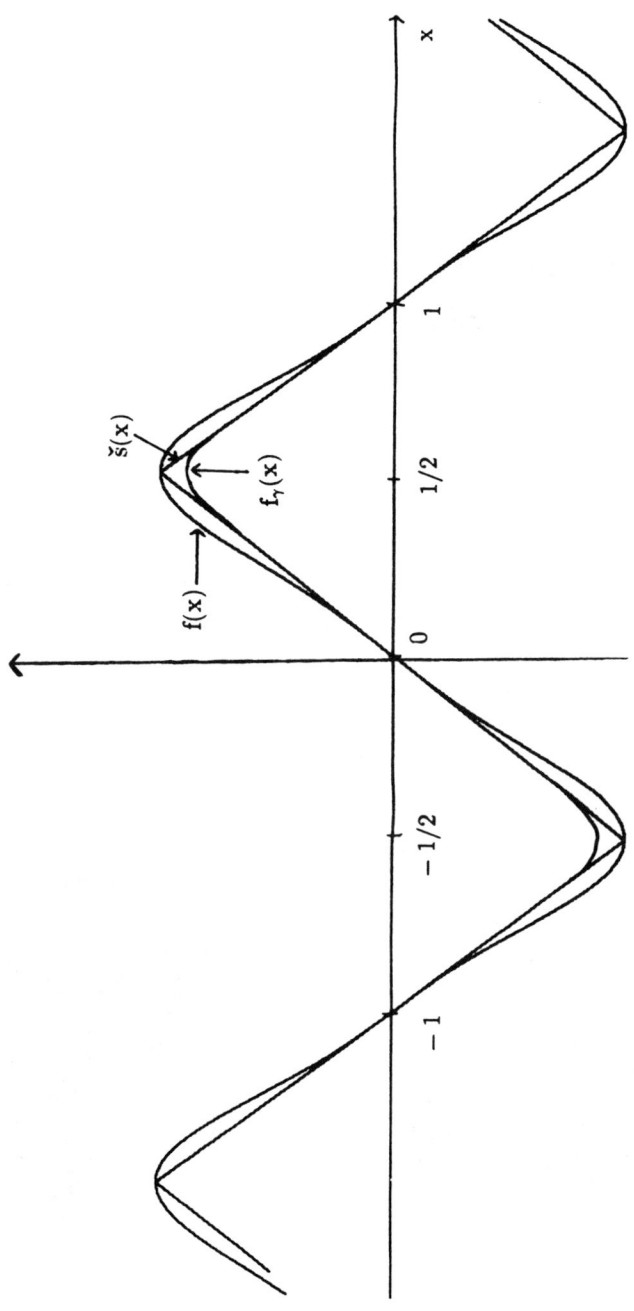

FIGURE 5. The nonlinearities $\check{s}$ and
$f(x) = (1/\pi)[\sin(\pi x) + (\pi/2 - 1)\sin^3(\pi x)]$, and f_γ ($\gamma > 0$)

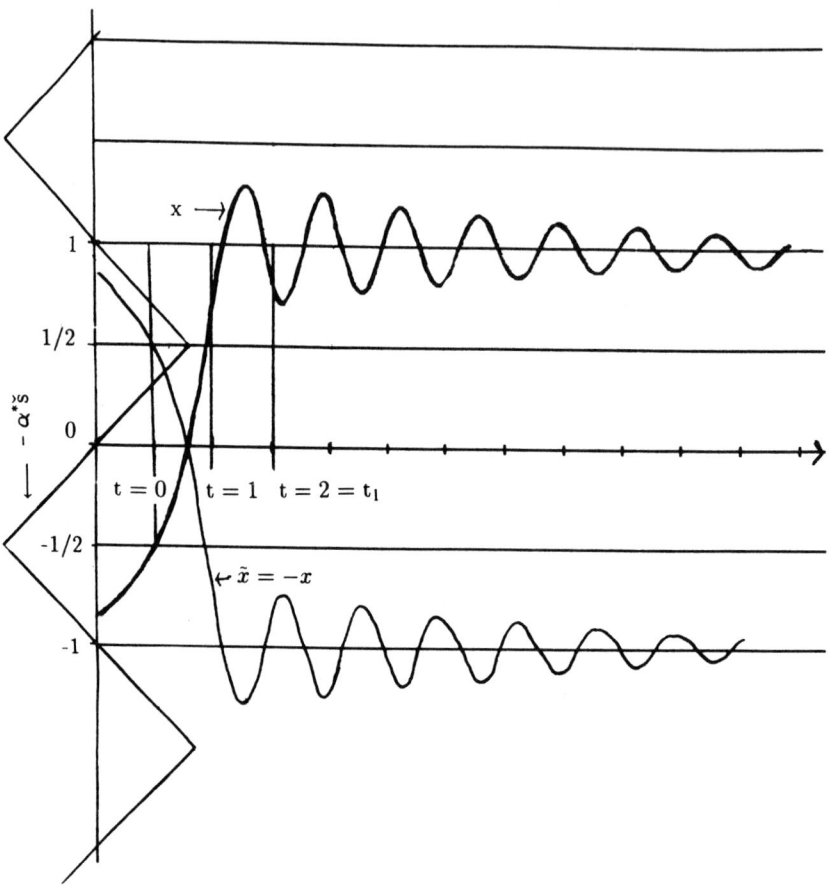

FIGURE 6. The heteroclinic solutions x and $\tilde{x} = -x$ of the piecewise linear equation $(-\alpha^*\check{s})$.
(The graph of $(-\alpha^*\check{s})$ is drawn vertically.)

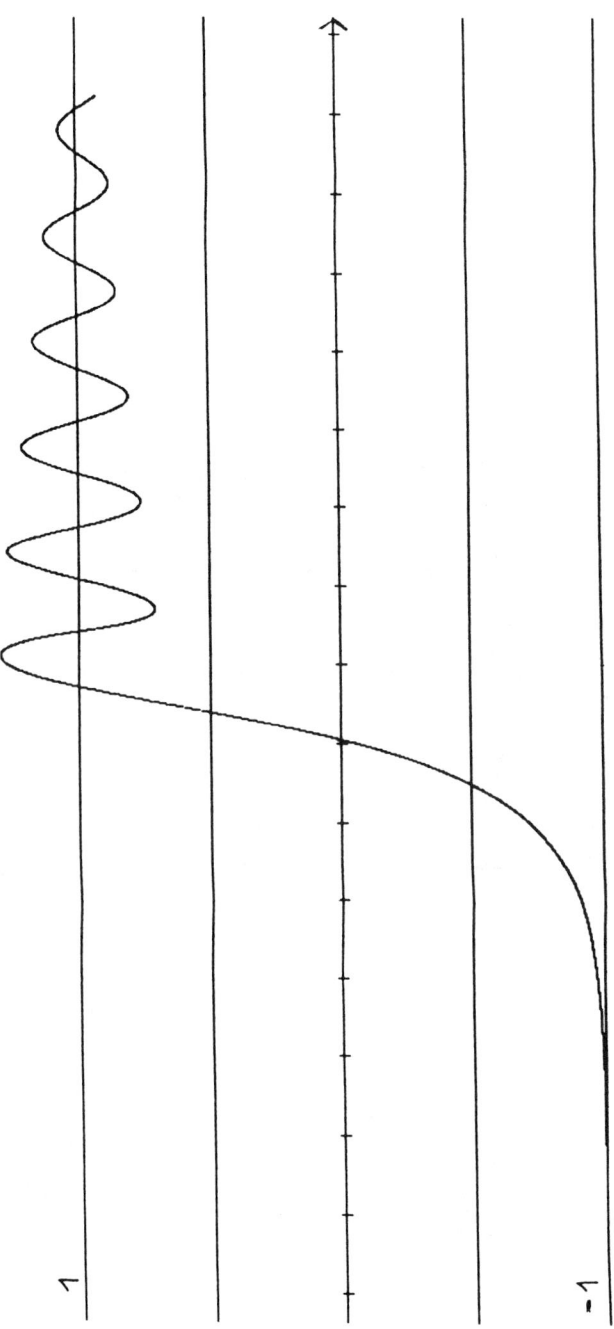

FIGURE 7. Numerically obtained 'heteroclinic' solution of
$\dot{x}(t) = -\alpha f(x(t-1))$, with
$f(x) = (1/\pi)[\sin(\pi x) + (\pi/2 - 1)\sin^3(\pi x)]$, and $\alpha \approx 3.526...$

BERNHARD LANI–WAYDA
MATHEMATISCHES INSTITUT DER UNIVERSITÄT GIESSEN
ARNDTSTR. 2
35392 GIESSEN
GERMANY
 E-mail address: Bernhard.Lani-Wayda@math.uni-giessen.de

Editorial Information

To be published in the *Memoirs*, a paper must be correct, new, nontrivial, and significant. Further, it must be well written and of interest to a substantial number of mathematicians. Piecemeal results, such as an inconclusive step toward an unproved major theorem or a minor variation on a known result, are in general not acceptable for publication. Papers appearing in *Memoirs* are generally longer than those appearing in *Transactions*, which shares the same editorial committee.

As of January 31, 2001, the backlog for this journal was approximately 7 volumes. This estimate is the result of dividing the number of manuscripts for this journal in the Providence office that have not yet gone to the printer on the above date by the average number of monographs per volume over the previous twelve months, reduced by the number of volumes published in four months (the time necessary for preparing a volume for the printer). (There are 6 volumes per year, each containing at least 4 numbers.)

A Consent to Publish and Copyright Agreement is required before a paper will be published in the *Memoirs*. After a paper is accepted for publication, the Providence office will send a Consent to Publish and Copyright Agreement to all authors of the paper. By submitting a paper to the *Memoirs*, authors certify that the results have not been submitted to nor are they under consideration for publication by another journal, conference proceedings, or similar publication.

Information for Authors

Memoirs are printed from camera copy fully prepared by the author. This means that the finished book will look exactly like the copy submitted.

The paper must contain a *descriptive title* and an *abstract* that summarizes the article in language suitable for workers in the general field (algebra, analysis, etc.). The *descriptive title* should be short, but informative; useless or vague phrases such as "some remarks about" or "concerning" should be avoided. The *abstract* should be at least one complete sentence, and at most 300 words. Included with the footnotes to the paper should be the 2000 *Mathematics Subject Classification* representing the primary and secondary subjects of the article. The classifications are accessible from www.ams.org/msc/. The list of classifications is also available in print starting with the 1999 annual index of *Mathematical Reviews*. The Mathematics Subject Classification footnote may be followed by a list of *key words and phrases* describing the subject matter of the article and taken from it. Journal abbreviations used in bibliographies are listed in the latest *Mathematical Reviews* annual index. The series abbreviations are also accessible from www.ams.org/publications/. To help in preparing and verifying references, the AMS offers MR Lookup, a Reference Tool for Linking, at www.ams.org/mrlookup/. When the manuscript is submitted, authors should supply the editor with electronic addresses if available. These will be printed after the postal address at the end of the article.

Electronically prepared manuscripts. The AMS encourages electronically prepared manuscripts, with a strong preference for $\mathcal{AMS}$-LaTeX. To this end, the Society has prepared $\mathcal{AMS}$-LaTeX author packages for each AMS publication. Author packages include instructions for preparing electronic manuscripts, the *AMS Author Handbook*, samples, and a style file that generates the particular design specifications of that publication series. Though $\mathcal{AMS}$-LaTeX is the highly preferred format of TeX, author packages are also available in $\mathcal{AMS}$-TeX.

Authors may retrieve an author package from e-MATH starting from `www.ams.org/tex/` or via FTP to `ftp.ams.org` (login as `anonymous`, enter username as password, and type `cd pub/author-info`). The *AMS Author Handbook* and the *Instruction Manual* are available in PDF format following the author packages link from `www.ams.org/tex/`. The author package can be obtained free of charge by sending email to `pub@ams.org` (Internet) or from the Publication Division, American Mathematical Society, P.O. Box 6248, Providence, RI 02940-6248. When requesting an author package, please specify $\mathcal{AMS}$-LaTeX or $\mathcal{AMS}$-TeX, Macintosh or IBM (3.5) format, and the publication in which your paper will appear. Please be sure to include your complete mailing address.

Sending electronic files. After acceptance, the source file(s) should be sent to the Providence office (this includes any TeX source file, any graphics files, and the DVI or PostScript file).

Before sending the source file, be sure you have proofread your paper carefully. The files you send must be the EXACT files used to generate the proof copy that was accepted for publication. For all publications, authors are required to send a printed copy of their paper, which exactly matches the copy approved for publication, along with any graphics that will appear in the paper.

TeX files may be submitted by email, FTP, or on diskette. The DVI file(s) and PostScript files should be submitted only by FTP or on diskette unless they are encoded properly to submit through email. (DVI files are binary and PostScript files tend to be very large.)

Electronically prepared manuscripts can be sent via email to `pub-submit@ams.org` (Internet). The subject line of the message should include the publication code to identify it as a Memoir. TeX source files, DVI files, and PostScript files can be transferred over the Internet by FTP to the Internet node `e-math.ams.org` (130.44.1.100).

Electronic graphics. Comprehensive instructions on preparing graphics are available at `www.ams.org/jourhtml/graphics.html`. A few of the major requirements are given here.

Submit files for graphics as EPS (Encapsulated PostScript) files. This includes graphics originated via a graphics application as well as scanned photographs or other computer-generated images. If this is not possible, TIFF files are acceptable as long as they can be opened in Adobe Photoshop or Illustrator. No matter what method was used to produce the graphic, it is necessary to provide a paper copy to the AMS.

Authors using graphics packages for the creation of electronic art should also avoid the use of any lines thinner than 0.5 points in width. Many graphics packages allow the user to specify a "hairline" for a very thin line. Hairlines often look acceptable when proofed on a typical laser printer. However, when produced on a high-resolution laser imagesetter, hairlines become nearly invisible and will be lost entirely in the final printing process.

Screens should be set to values between 15% and 85%. Screens which fall outside of this range are too light or too dark to print correctly. Variations of screens within a graphic should be no less than 10%.

Inquiries. Any inquiries concerning a paper that has been accepted for publication should be sent directly to the Electronic Prepress Department, American Mathematical Society, P. O. Box 6248, Providence, RI 02940-6248.

Editors

This journal is designed particularly for long research papers, normally at least 80 pages in length, and groups of cognate papers in pure and applied mathematics. Papers intended for publication in the *Memoirs* should be addressed to one of the following editors. In principle the Memoirs welcomes electronic submissions, and some of the editors, those whose names appear below with an asterisk (*), have indicated that they prefer them. However, editors reserve the right to request hard copies after papers have been submitted electronically. Authors are advised to make preliminary email inquiries to editors about whether they are likely to be able to handle submissions in a particular electronic form.

Algebra to CHARLES CURTIS, Department of Mathematics, University of Oregon, Eugene, OR 97403-1222 email: cwc@darkwing.uoregon.edu

Algebraic geometry and commutative algebra to LAWRENCE EIN, Department of Mathematics, University of Illinois, 851 S. Morgan (M/C 249), Chicago, IL 60607-7045; email: ein@uic.edu

Algebraic topology and cohomology of groups to STEWART PRIDDY, Department of Mathematics, Northwestern University, 2033 Sheridan Road, Evanston, IL 60208-2730; email: priddy@math.nwu.edu

Combinatorics and Lie theory to SERGEY FOMIN, Department of Mathematics, University of Michigan, Ann Arbor, Michigan 48109-1109; email: fomin@math.lsa.umich.edu

Complex analysis and complex geometry to DUONG H. PHONG, Department of Mathematics, Columbia University, 2990 Broadway, New York, NY 10027-0029; email: dp@math.columbia.edu

*****Differential geometry and global analysis** to LISA C. JEFFREY, Department of Mathematics, University of Toronto, 100 St. George St., Toronto, ON Canada M5S 3G3; email: jeffrey@math.toronto.edu

*****Dynamical systems and ergodic theory** to ROBERT F. WILLIAMS, Department of Mathematics, University of Texas, Austin, Texas 78712-1082; email: bob@math.utexas.edu

Geometric topology, knot theory and hyperbolic geometry to ABIGAIL A. THOMPSON, Department of Mathematics, University of California, Davis, Davis, CA 95616-5224; email: thompson@math.ucdavis.edu

Harmonic analysis, representation theory, and Lie theory to ROBERT J. STANTON, Department of Mathematics, The Ohio State University, 231 West 18th Avenue, Columbus, OH 43210-1174; email: stanton@math.ohio-state.edu

*****Logic** to THEODORE SLAMAN, Department of Mathematics, University of California, Berkeley, CA 94720-3840; email: slaman@math.berkeley.edu

Number theory to MICHAEL J. LARSEN, Department of Mathematics, Indiana University, Bloomington, IN 47405; email: larsen@math.indiana.edu

Operator algebras and functional analysis to BRUCE E. BLACKADAR, Department of Mathematics, University of Nevada, Reno, NV 89557; email: bruceb@math.unr.edu

*****Ordinary differential equations, partial differential equations, and applied mathematics** to PETER W. BATES, Department of Mathematics, Brigham Young University, 292 TMCB, Provo, UT 84602-1001; email: peter@math.byu.edu

*****Partial differential equations and applied mathematics** to BARBARA LEE KEYFITZ, Department of Mathematics, University of Houston, 4800 Calhoun Road, Houston, TX 77204-3476; email: keyfitz@uh.edu

*****Probability and statistics** to KRZYSZTOF BURDZY, Department of Mathematics, University of Washington, Box 354350, Seattle, Washington 98195-4350; email: burdzy@math.washington.edu

*****Real and harmonic analysis and geometric partial differential equations** to WILLIAM BECKNER, Department of Mathematics, University of Texas, Austin, TX 78712-1082; email: beckner@math.utexas.edu

All other communications to the editors should be addressed to the Managing Editor, WILLIAM BECKNER, Department of Mathematics, University of Texas, Austin, TX 78712-1082; email: beckner@math.utexas.edu.

Selected Titles in This Series

(*Continued from the front of this publication*)

688 **Piotr Hajłasz and Pekka Koskela,** Sobolev met Poincaré, 2000

687 **Guy David and Stephen Semmes,** Uniform rectifiability and quasiminimizing sets of arbitrary codimension, 2000

686 **L. Gaunce Lewis, Jr.,** Splitting theorems for certain equivariant spectra, 2000

685 **Jean-Luc Joly, Guy Metivier, and Jeffrey Rauch,** Caustics for dissipative semilinear oscillations, 2000

684 **Harvey I. Blau, Bangteng Xu, Z. Arad, E. Fisman, V. Miloslavsky, and M. Muzychuk,** Homogeneous integral table algebras of degree three: A trilogy, 2000

683 **Serge Bouc,** Non-additive exact functors and tensor induction for Mackey functors, 2000

682 **Martin Majewski,** ational homotopical models and uniqueness, 2000

681 **David P. Blecher, Paul S. Muhly, and Vern I. Paulsen,** Categories of operator modules (Morita equivalence and projective modules, 2000

680 **Joachim Zacharias,** Continuous tensor products and Arveson's spectral C^*-algebras, 2000

679 **Y. A. Abramovich and A. K. Kitover,** Inverses of disjointness preserving operators, 2000

678 **Wilhelm Stannat,** The theory of generalized Dirichlet forms and its applications in analysis and stochastics, 1999

677 **Volodymyr V. Lyubashenko,** Squared Hopf algebras, 1999

676 **S. Strelitz,** Asymptotics for solutions of linear differential equations having turning points with applications, 1999

675 **Michael B. Marcus and Jay Rosen,** Renormalized self-intersection local times and Wick power chaos processes, 1999

674 **R. Lawther and D. M. Testerman,** A_1 subgroups of exceptional algebraic groups, 1999

673 **John Lott,** Diffeomorphisms and noncommutative analytic torsion, 1999

672 **Yael Karshon,** Periodic Hamiltonian flows on four dimensional manifolds, 1999

671 **Andrzej Rosłanowski and Saharon Shelah,** Norms on possibilities I: Forcing with trees and creatures, 1999

670 **Steve Jackson,** A computation of δ_5^1, 1999

669 **Seán Keel and James McKernan,** Rational curves on quasi-projective surfaces, 1999

668 **E. N. Dancer and P. Poláčik,** Realization of vector fields and dynamics of spatially homogeneous parabolic equations, 1999

667 **Ethan Akin,** Simplicial dynamical systems, 1999

666 **Mark Hovey and Neil P. Strickland,** Morava K-theories and localisation, 1999

665 **George Lawrence Ashline,** The defect relation of meromorphic maps on parabolic manifolds, 1999

664 **Xia Chen,** Limit theorems for functionals of ergodic Markov chains with general state space, 1999

663 **Ola Bratteli and Palle E. T. Jorgensen,** Iterated function systems and permutation representation of the Cuntz algebra, 1999

662 **B. H. Bowditch,** Treelike structures arising from continua and convergence groups, 1999

661 **J. P. C. Greenlees,** Rational S^1-equivariant stable homotopy theory, 1999

660 **Dale E. Alspach,** Tensor products and independent sums of $\mathcal{L}_p$-spaces, $1 < p < \infty$, 1999

659 **R. D. Nussbaum and S. M. Verduyn Lunel,** Generalizations of the Perron-Frobenius theorem for nonlinear maps, 1999

For a complete list of titles in this series, visit the
AMS Bookstore at **www.ams.org/bookstore/**.